Advanced Sciences and Technologies for Security Applications

Indexed by SCOPUS

The series Advanced Sciences and Technologies for Security Applications comprises interdisciplinary research covering the theory, foundations and domain-specific topics pertaining to security. Publications within the series are peer-reviewed monographs and edited works in the areas of:

- biological and chemical threat recognition and detection (e.g., biosensors, aerosols, forensics)
- crisis and disaster management
- terrorism
- cyber security and secure information systems (e.g., encryption, optical and photonic systems)
- traditional and non-traditional security
- energy, food and resource security
- economic security and securitization (including associated infrastructures)
- transnational crime
- human security and health security
- social, political and psychological aspects of security
- recognition and identification (e.g., optical imaging, biometrics, authentication and verification)
- smart surveillance systems
- applications of theoretical frameworks and methodologies (e.g., grounded theory, complexity, network sciences, modelling and simulation)

Together, the high-quality contributions to this series provide a cross-disciplinary overview of forefront research endeavours aiming to make the world a safer place.

The editors encourage prospective authors to correspond with them in advance of submitting a manuscript. Submission of manuscripts should be made to the Editor-in-Chief or one of the Editors.

More information about this series at https://link.springer.com/bookseries/5540

Reza Montasari · Hamid Jahankhani ·
Haider Al-Khateeb

Editors

Challenges in the IoT and Smart Environments

A Practitioners' Guide to Security, Ethics and Criminal Threats

 Springer

Editors
Reza Montasari
Hillary Rodham Clinton School of Law
Swansea University
Swansea, UK

Hamid Jahankhani
Northumbria University
London, UK

Haider Al-Khateeb
School of Mathematics and Computer
Science
University of Wolverhampton
Wolverhampton, UK

ISSN 1613-5113   ISSN 2363-9466 (electronic)
Advanced Sciences and Technologies for Security Applications
ISBN 978-3-030-87168-0   ISBN 978-3-030-87166-6 (eBook)
https://doi.org/10.1007/978-3-030-87166-6

This Springer imprint is published by the registered company Springer Nature Switzerland AG
The registered company address is: Gewerbestrasse 11, 6330 Cham, Switzerland

Contents

Security in Network Services Delivery for 5G Enabled D2D Communications: Challenges and Solutions

Ed Kamya Kiyemba Edris⬤, Mahdi Aiash⬤, and Jonathan Loo⬤

Abstract Due to the increasing data traffic in mobile network, fifth generation (5G) will use Device to Device (D2D) communications as the underlay technology to offload traffic from the 5G core network (5GC) and push content to the edge closer to the users by taking advantage of proximity and storage features of User Equipment (UE). This will be supported by Networks Services (NS) provided by 5G, it will also enable new use cases that will provide the accessibility to other services in the Home network (HN) and third-party service provider (SP). Mobile Network Operators (MNO) will use NS such as D2D communications and content-centric networking (CCN) to deliver content-based services efficiently to UE. Both D2D and CCN have known security issues and their integration brings new security challenges. In this article, we present an integrated network service delivery (NSD) framework for 5G enabled D2D communications that leverages on NS for service discovery, content delivery, and protection of data. We also present a comprehensive investigation of security and privacy in D2D communications and service-oriented network, highlighting the vulnerabilities and threats on the network. We also evaluate the security requirements of NSD to deliver NS securely to D2D users based on X.805 security framework using the eight security dimensions and level abstraction approach for a systematic and comprehensive approach. Finally, we recommend security solution approaches for the secure NS access and sharing of data between users in 5G enabled D2D communications network.

Supported by organization x.

E. K. K. Edris · M. Aiash (✉)
Middlesex University, London, UK
e-mail: m.aiash@mdx.ac.uk

E. K. K. Edris
e-mail: ee351@live.mdx.ac.uk

J. Loo
University of West London, London, UK
e-mail: jonathan.loo@uwl.ac.uk

Keywords 5G · Survey · Network services · Security analysis · Content sharing · Service delivery · Device-to-device · Content-centric · X.805 framework

1 Introduction

Over the past few years there has been an increase in the mobile traffic due to rising demand of over-the-top applications (OTT) such as social media, live streaming, local based advertising, and popularity of smart phones usage [15]. Mobile traffic is anticipated to keep growing gradually, most of traffic is from mobile devices and Machine to Machine communications (M2M) [17]. The demand of multimedia contents and expectation of high availability and performance by the end users affected the network capacity. Whereby network infrastructures were overloaded and became highly inefficient for content distribution [38]. Most of the data traffic load of the mobile network backhaul is generated from the mobile user's traffic. The fifth-generation mobile network (5G) was inspired by the need for very high reliability, ultra-low latency network to support services and the increasing demand of quick access and delivery of data by end users and Mobile Network Operators (MNO) [30]. 5G will enable the end users to be involved in content-based operations through Device to Device (D2D) communications, which will enhance user's experience. D2D Communications will be used as an underlay technology for 5G to offload traffic from the network backhaul by pushing content to edge closer to the end users [31]. D2D communications will also be fundamental to the implementation of Internet of Things (IoT). Content distribution and retrieval will dominate mobile traffic, however delivering such services to the end user efficiently and securely is becoming a big challenge.

D2D communications [2] was specified by the Third Generation Partnership Project (3GPP) with a purpose of network offloading via the User Equipment (UE) by communicating directly without conveying content through the Base Station (BS). In 5G, the UE will act as a data consumer as well as playing a role in content distribution and delivery [1]. 5G enabled D2D communications will support new use cases and services, the delivery of these services to the end user will be facilitated by Network Services (NS) by using context aware enabled devices. To be able to distribute and deliver contents to the end user various content delivery models such as Content Delivery Network (CDN) [38] will be used. CDN can be deployed at edge, BS, and access points to support cache servers and with D2D communications, mobile devices can also be used as cache nodes [28]. Most of the wireless traffic generated in cellular network is from the downloads of popular content replicated in multiple locations [47]. The introduction of CDN and content caching in mobile network can be integrated with Information-Centric Networking (ICN), transforming the network from connection centric to information centric such as Content-Centric Networking (CCN) [36] in next generation mobile network [41].

Due heterogeneous nature of 5G, UE, network and the data will be vulnerable to new and old attacks when UEs are accessing services in the Home Network

(HN) and third-party service provider (SP). In addition, UEs sharing continent in out of coverage scenario without control of the network will need robust protection. So far, less attention has been given to the security issues faced by an integrated service framework for mobile network as the one presented in this article. To best of our knowledge no study has been carried out so far to investigate the security threats and requirements of Network Service Delivery (NSD) in 5G enabled D2D communications, using a systematic approach, moreover there is need for a security framework for NSD. The provision of secure NSD is essential to achieving the main objectives of 5G.

The study is motivated by a secure NSD in 5G enabled D2D communications leveraging on other NSs for service discovery, content delivery, and protection of data. The emphasis is on the protection of data and communication channels from different threats. To give a systematic approach of the security evaluation for 5G enabled D2D communications, we apply the X.805 security framework [73] with NS abstraction level approach [20]. This framework has been used to evaluate end to end security of communication centric systems such as 4G [55], internet of things [57], and ICN [49].

The article adds the following contributions: firstly, it presents a NSD framework for delivery and sharing services between UEs in different scenarios. Secondly, it comprehensively investigates of security and privacy of 5G enabled D2D communications network. It identifies the potential threats against D2D communications and ICN integrated system model. Thirdly, it evaluates the security requirements using a systematic and abstract approach for more comprehensive evaluation using X.805 framework. Lastly, suggests possible solutions of the threats and mention future work on NSD.

The rest of the article is organized as follows. Section II presents an overview of NSD framework for 5G enabled D2D communications architecture, system, and service models. Security threats and threat model are discussed in section III. While in Section IV, X.805 framework is used to evaluate the security requirements of 5G enabled D2D communications. The existing security solutions and new approaches are discussed in section V. Finally, the article is concluded in section VI.

2 Network Service Delivery Framework

To investigate security of NSD, an access and delivery framework is presented in this section based on the network architecture in [16], NS abstraction in [20], and CCN architecture. It focuses on the entities' communication and how the services are accessed, cached and shared between UEs but not how the data is stored or accessed on the application level.

2.1 Network Architecture

Since D2D and ProSe functionalities for 5G are in process of being standardised, this article uses the D2D communications architecture in [20] as shown in Fig. 1 and presents an NSD framework that utilize on mobile content delivery and network functions. CCN is integrated with cellular network to enable content aware operations such as content resolution at edge which is within 3GPP standardization [14]. 5G Network Function Virtualization (NFV) [43] allows the sharing of infrastructure resources between Network Service providers (NSP) and the delivery of content to users through network slicing. D2D communications, content delivery and content sharing are classified as NSs. ICN in 5G can be implemented using NFV and Software Defined Network (SDN) where ICN based service delivery methods such as CCN inherently integrate into the network infrastructure [58].

5G adopts C-RAN architecture [16], which centralizes the baseband resources to a single virtualized Base Band Unit (BBU) pool, then connects to several radio transceiver units called the Remote Radio Heads (RRHs) [39]. It enables virtualization, facilitating network infrastructure sharing [42] and the interface between BBU and RRH has been changed from circuit fronthaul to packet fronthaul. BBU is divided into Distributed Unit (DU) which is responsible for physical layer as well as real time Media Access Control (MAC) layer process and Centralized Unit (CU) responsible for upper layer computations process [66]. In 5G, the UE connects to 5G core network (5GC) via the new generation Radio Access Network (ngRAN) then to other Network Functions (NF) such as Access and Mobility Function (AMF), Session Management Function (SMF), Authentication Server Function (AUSF), User Plane Function (UPF), and Unified Data Management (UDM) as defined in [3], illustrated in Fig. 2.

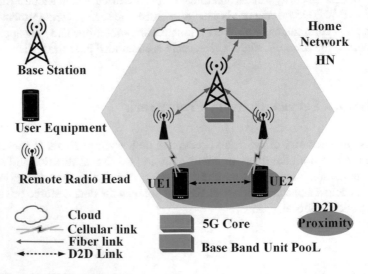

Fig. 1 5G enabled D2D architecture

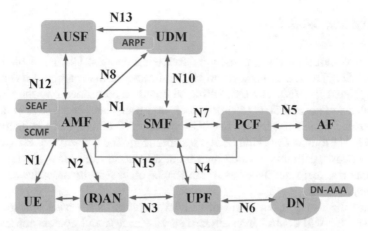

Fig. 2 5G system architecture

2.2 Security Architecture

The current security standards are not adequate for 5G, which affects new mission critical services and use cases. Therefore, there is a need for security architecture [4] that enforces trust between new actors and other entities in the HN [21]. The mobile network still consisted of three essential parties:

– UE: Consists of a Mobile Equipment (ME) and the Universal Subscriber Identity Module (USIM).
– Home Network (HN): It houses the user database and other security functions that stores users' subscription data and security credentials such as Subscription Permanent Identifier (SUPI) and the long-term key K.
– Serving Network (SN): The access network that the UE connects to via ngRAN.

The security architecture introduces new security entities such as Security Anchor Function (SEAF), AUSF, Authentication Credential Repository and Processing Function (ARPF) [4, 21]. The security architecture will have to consider NFV and SDN to achieve the objectives of 5G. Security enablers in 5G need to address key security concerns such as authentication, authorization, availability, privacy, trust, security monitoring, network management and virtualization isolation. Security Control Classes (SCC) are introduced to describe the security aspect of 5G system, SCC are mapped with security requirements based on X.805 eight security dimensions [5, 73].

2.3   System Model

The system model in Fig. 3 consists of following entities: UE, BBU pool, RRH, HN, and SN. The UE is registers to HN and receives the roaming services from Visiting Network (VN). The CCN protocol could be embedded into the UE, BBU pool, edge routers and 5GC [58] or control and user plane enhancement could be implemented to enable services like ICN within 5GC and extend the interfaces to support ICN Protocol Data Unit (PDU) sessions [59]. The UE will request to connect to the network, get authenticated then request to access the other services as per subscription agreements. To access these service other security procedures might be required [21–23].

Beside the usual management of cellular network and service operations, the MNO and SP will control internetwork service access and content retrieval to a certain extent. In 5G, the SP could be the MNO, third-party or another SP using the shared infrastructure as tenant provided by the MNO through network slicing. With MNO controlling service access and security management, it will be able hide its visibility or deny the UE from accessing a particular services. We more interested in content access and retrieval process part of D2D communication and discovery.

For data access and delivery, the UE requests for a content and interest message is forwarded to the BBU through RRH using the CCN forwarding process to full the UE request [58, 76]. The request could satisfied at by UE in proximity, Content Store (CS) in BBU Pool or in the 5GC CS. The CCN process involves local caching and satisfaction at different levels of the system model like the at edge as shown in Fig. 3. This enables offloading of the traffic from the backhaul to the edge if the request and data matching do not reach the 5GC. BBU also discovers cached

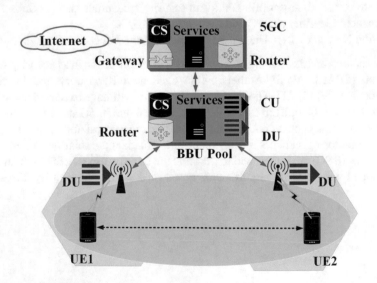

Fig. 3   System model

content by associated devices and content transmission is performed through D2D communications [37]. However this process exposes the involved entities and data to various threats, hence the need for a comprehensive security requirements evaluation for the system model.

3 Security and Privacy for NSD in 5G Enabled D2D Communications

In this section, the threat model and threats that can affect the system model are presented. Security and privacy are serious concerns in 5G enabled D2D communications due to its characteristics. The UE's participation in content transmission, distribution, delivery and traffic offloading [33], not only increases its own exposure to threats but also of other entities and the data. Mobile security at edge is another concern, where the services and user's data will be most accessed. Also, communication channels between networks and D2D devices will be vulnerable to attacks, even the HN and VN might want to eavesdrop on D2D communications [68]. The MNO must authenticate and validate the SP to ensure legitimate access and provision of their services. In addition, the security context could be compromised and exposed outside the HN [5].

3.1 Threat Model

To evaluate security and privacy protection mechanisms for 5G enabled D2D communications, it requires a clear adversary. In this case , the threat model used is based on a Dolev–Yao (DY) model [19], an adversary model that formally models the attack against communication. DY is assumed to be the communication channel, capable of tapping the channel and eavesdrop on the transmitted messages. The DY can create, read, capture, replay and send messages on the wireless and wired communication channel in the network. In addition, the adversary can compromise UEs by revealing the secrets between UEs as well as applying her own public functions such as encryption and hashing. The adversary can impersonate any entity participating in the transaction, capable of initiating communication and responding to interest message sent by legitimate UEs. Might also try to repudiate their malicious behavior, preventing data sharing between UEs, hence denying service to other UEs and network entities.

3.2 Security Threats

Most studies have focussed on ICN integration with mobile network architecture, caching strategies [15, 38, 67]. Other studies have investigated security in ICN and D2D communications separately, while security issues in an integrated mobile network have not been investigated. The security and privacy in D2D communications were investigated in [34, 68, 75] however they did not use a systematic approach in their investigations. While the authors in [49] investigated the challenges in ICN based on NetInf architecture, evaluated the security requirements using the X.805 framework. They described the problems without the scope of the future challenges. In [65] a survey was conducted on ICN security but did not elaborate on the security challenges in heterogeneous networks such as next generation mobile network. While [7] conducted a survey on attacks affecting most ICN architectures but did not include attacks on CCN.

The wireless nature of D2D communications plus its characteristics and architecture present several security vulnerabilities that put network at risk to potential threats [1]. 5G can still be affected by the vulnerabilities from legacy systems and the security of NS might be compromised by new attacks on different levels of the network including the network slices. In cellular network, the BS acts as a Central Authority (CA) but in D2D communications it might play a minimal part hence strong anonymity is provided. However, the BS will have access to the data transmitted between the UE and other UEs, which could expose the data to possible attacks in form of active and passive, local and extended. We discuss some of these attacks below:

Eavesdropping: D2D messages can be eavesdropped by unauthorized users and authorized cellular users. In addition, side channel attacks across network slices targeting the implementation of cryptography or running a code to influence the contents of the cache [10].

Data Fabrication: The unprotected transmitted data can be fabricated or changed by malicious users, which leads to the content being circulated by unaware infected device to other devices such as modification of control data.

Impersonation Attack: A legitimate user might be impersonated by malicious user and communicate to other D2D users through identity impersonation and masquerading attacks. Also, network slice instance are vulnerable to these attack, which could lead to other attacks like monitoring or location attacks [10].

Free-Riding Attack: In D2D communications devices participate in sending and receiving data willingly but some UEs might not be willing to send data to others when in power saving mode while receiving data from its peers, which decreases the system availability.

Privacy Violation: It is important to protect the privacy of users' data such as their identity and location. If an attacker is able to listen and intercept transmitted messages and she would be able to extract information and guess the location of the UE. ICN cached content, user privacy and content names are all targets of privacy attacks. Moreover, user's subscription information could be leaked by a malicious attacker or compromised publisher through attacks such as timing, protocol and anonymity attacks [65].

Denial of Service (DoS) Attack: D2D services might be interrupted by making them unavailable to the intended users, by weakening or blocking legitimate devices from establishing connection completely. An attacker can send big continuous request to ICN nodes for services such as content, domain name queries or initiate interest flooding attack [6]. Also an attacker might exhaust security resources in one network slice so that she can attacker other slices [10].

Content Poisoning Attack: It involves filling the content router's cache with invalid content that ahs valid name matching the sent interest, however, the payload might be fake or with invalid signature [26].

Cache Pollution Attack: A malicious attacker may weaken caching activity by requesting less popular content frequently with attacks such as locality disruption and false locality [65].

Unauthorized access: An unauthorized node might access an object which was intended for a specific entity. For instance, in unauthorized cache access, a cached object from a local device might be access by unauthorized device [49].

Cache Misuse: The attacker can utilize on caches capability and use it as storage, hence, enabling the attacker to make her own content available. Also, an attacker can corrupt the cache content turning it into incorrect returned objects for DoS attack.

False Accusation: A malicious publisher tries to make it look like the requester has requested an object when it is not the case and might also charge a subscriber for services that was never requested or obtained.

IP Spoofing: Attackers uses malicious code to manipulate header of IP packets.

Location Spoofing: Attacker sends a fake location information to disturb D2D formation by imitating with artificial locations to confuse the D2D members.

Session Hijacking: The attacker spoofs the IP address of the victim device and guesses the sequence number expected by the targeted source device, this is followed by a Distributed DoS (DDoS) attack on the victim device, impersonate the device to carry on the session with the targeted device.

Communication Monitoring: The attacker with access to the same router as that requester is using to receive content then the attacker targets a requester and tries to identify the victim's requested contents.

Jamming Attack: Malicious user masquerading as legitimate subscriber sends many malicious content requests to disrupt the flow of information and replies are sent to a destination other than the requester's. In 5G jamming is achieved through analysing physical control link channels and signals [44].

Data Leakage: The UE might be attached to several slices on the network level with different security parameters. If the UE cannot separate data from different slices, the separation between slices could decrease, leading to the UE receiving sensitive data from one slice and then publish that data via another slice [10].

4   Security Evaluation of NSD in 5G Enabled D2D Communications Using X.805 Framework

In this section, X.805 framework [73] is applied to evaluate the security requirements for delivering secure NS based on a system model, threat model and the NS abstraction in [20] which are mapped with X.805 security layers, planes and dimensions.

4.1   The X.805 Security Framework Overview

Evaluating security in any networking system is very complicated, International Telecommunication Union Standardization Sector (ITU-T) developed X.805 framework as a security analysis tool. The X.805 framework uses a modular method to create a multi layered framework which assesses possible threats and vulnerabilities in end to end security to address security threats in networking systems effectively. The X.805 defines following: three security layers (applications, services and infrastructure); three security planes (end user, control and management; and eight security dimensions (access control, authentication, non- reputation, data confidentiality, communication security, data integrity, availability and privacy). The security layers and security planes are identified according to the network activities as illustrated in Fig. 4. In addition, nine security viewpoints are created by applying each security plane to each security layer, whereby each viewpoint has its own distinctive vulnerabilities and threats based on security dimensions.

Security Layers: The infrastructure security layer covers the fundamental building blocks of NS, NF, network slices, applications and individual communication links such as BS, RRH, routers, servers, slices and fibre links. This layer facilitates security of hosts involved in the data transmission, it prevents attacks from air interfaces

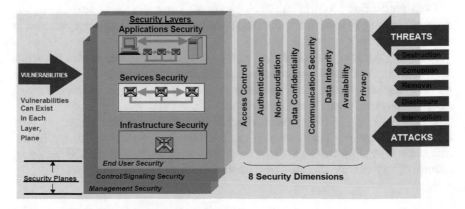

Fig. 4 X.805 security framework (Zeltsan, 2005)

Table 1 Infrastructure layer in relations with security dimensions

Security dimensions	Infrastructure layer	Security mechanisms
Access Control	Authorize UE and network entities to accessing data on the UE and other entities	ACL, passwords
Authentication	Verify the identity of the UE, BBU and server providing the NS to the UE	Shared secret, PKI digital signature, digital certificate
Non-repudiation	Record UE, BBU, servers that perform activities on devices while accessing the NS	MAC, hash , function encryption
Data confidentiality	Protect the data on network devices, on UE and and control data	Symmetric and asymmetric encryption
Communication security	Ensures that UE, control and management data is only transmitted on sure channels	Symmetric and asymmetric encryption
Data integrity	Protect data on network entities, in transit and control data against unauthorized modification	MAC, hash function, digital signature
Availability	Ensure that network devices can receive and access UE data and manage D2D links	IDS,IPS, BC, DR
Privacy	Ensure that data which can identify the entities is not available to an unauthorized users	Encryption

and physical links including content servers, gateways, BBU and D2D connectivity. While service security layer covers services provided to end-users such as CCN, IP, cellular, QoS and location services. Securing this layer is complicated by the fact that services may build-upon one another to satisfy user requirements such as sharing and delivery of services via D2D communications. The CCN is related to the service layer while D2D communications is related to the infrastructure layer.

Security Planes: The security planes are concerned with securing the operations and provisioning of the individual mobile network elements, communication link as well as securing the functions of NS such as the configuration of UE, BBU, 5GC and secure content provisioning. In addition, its concerned with securing the control data in the network elements and in transit for NS such as D2D control link and PDU session. It ensures the security of the end user data on the network elements.

Security Dimensions: The eight security dimensions are used as a viewpoint for vulnerabilities and threats to provide protection against any attack in form security controls such as the authentication, availability, integrity, confidentiality, and access control on each layer.

Table 2 Service layer in relations with security dimensions

Security dimensions	Service layer	Security mechanisms
Access Control	Authorize BBU and SP to perform management activities on NSs	ACL, passwords
Authentication	Verify the identity of the NSs, the service entities on the and the origin of the NS	Shared secret, PKI, digital signature digital certificate
Non-repudiation	Record the SP, UE, BBU to prevent deny the transactions and origin of the control message	MAC, hash function encryption
Data confidentiality	Protect the NS's data transiting the network devices from unauthorized access	Encryption and
Communication security	Ensures that NS management, control data for UE passes through a secure channel	Encryption
Data integrity	Protect the management, control, the UE data against unauthorized modification and deletion	MAC, hash function, digital signature
Availability	Ensure that the network devices UE data and D2D link available to receive control data	IDS,IPS, BC, DR
Privacy	Ensure that data that can be used to identify NS is not available to unauthorized users	Encryption

4.2 Security Evaluation Using X.805 Framework

To able to mitigate these potential threats against the system model, the security requirements must be evaluated. This article focusses on the service and infrastructure layers of the system. It classifies security requirements using security layers, associated with the security planes in modular format, each module is analysed using the eight security dimensions, summarised in Tables 1 and 2. The modules 1, 2 and 3 are based on infrastructure layer whereas modules 4, 5 and 6 are associated with service layer. The X.805 demonstrates a methodical approach in tabular form as shown in Figs. 5 and 6, the security objectives of the dimensions for each module are analysed. The security goal is to cover the security capability of the framework including the detection and recognition of attacks, protection of the system, audit of the system and its recovery after the attack. Based on the above potential threats, the NS using

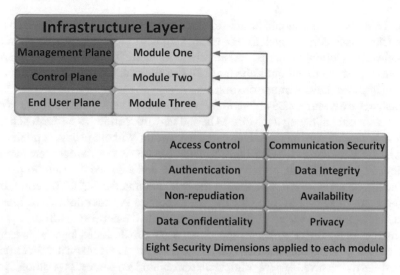

Fig. 5 Infrastructure layer with eight dimensions

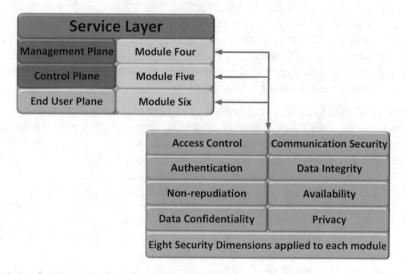

Fig. 6 Service layer with eight dimensions

infrastructure of cellular network should meet the certain security requirements. A comprehensive security analysis follows in the next subsections.

Access Control This dimension limits and control access to network elements and services through Access Control (ACL), encryption and authorization mechanisms. Some services lack a built in support to provide ACL or authorization framework. When an entity that is not controlled by the SP publishes content, the SP has no way of applying access control or knowing which user has accessed or cached data [53].

In addition, the system should be able to revoke user's privilege if it is detected to be a malicious user. Attacks such as free riding should be prevented and UE should be protected from joining rouge BS. Also, privileges of D2D users should be deprived in time if a user is found out to be malicious or their subscription has expired and revocability can prevent impersonation attack.

Infrastructure Security Layer—*Modules 1, 2 and 3*: The ACL at this layer is concerned with only allowing authorized UE and network entities to perform activities such as accessing data on the UE and in the network. Without the appropriate ACL policies, an unauthorized device might be able to access services which were intended for limited UEs. ACL can permit or deny the SP and a device the right to perform any action on service or device during the D2D communications. Therefore, extra mechanisms should be in place for the UE to control the data flow. ACL must be applied to control the access of data and activities on the network entities.

Service Security Layer— *Modules 4, 5 and 6*: The ACL at this layer is concerned with allowing only authorized BBU and SP to perform management activities on the NS and that the received service originated from authorized source. In addition, it only allows authorized UE to access the NS and to ensure the request message originated from an authorized UE before being accepting. The BBU should be able hide its services or visibility from the unauthorized UE after the authentication and during handover session. ACL mechanisms such as Role-Based Access Control (RBAC) or Discretionary Access Control (DAC) must be applied on the service entities and should depend on the identity of the subject for authorization, it suits unstructured domains like the system model in this article.

Authentication This dimension ensures a valid proof of identity is presented in form of shared secret, Public Key Infrastructure (PKI), digital certificate [8]. Authentication evaluates the identity of a party and verifies if the party is in possession of a secret and a key, it can be applied on an entity and data. Assigning an identity to a secret or key is required during authentication. For the UE to access the NS, the UE and network must perform authentication using Authentication and Key Agreement (AKA) methods [4]. In 5G, after a successful primary authentication, a secondary authentication can be performed to ensure that only authorized UEs can access SP services. UEs, service and the network must be able to mutually authenticate to stop attacks such as impersonation attack, false content injection and free riding.

Infrastructure Security Layer— *Modules 1, 2 and 3*: Authentication at this layer is concerned with verifying the identity of the UE requesting the services, SN, BBU and service entities providing the NS to the UE. The identity of the entities and transmitted data must be authenticated to secure D2D communications. The first step is to authenticate the entities to confirm the D2D peers' identities then authenticate data sources to confirm if it is from legitimate users [75]. Also, authentication can be achieved when the requester verifies signature that was used to digitally sign the data by the publisher. If UE A receives a message from UE B, A can verify that B is indeed the sender of the message where A and B can be any of the device in the network such as UE, server, which can also be classified as subscribers and MNO/SP [49].

Service Security Layer—*Modules 4, 5 and 6*: The authentication on this layer is concerned with verifying the identity of services and the origin of the NS. The verification of UE trying to access the services should be done by Authentication, Authorization and Accounting servers (3As), which also monitor, manage the subscription and service provisioned for the UE. The service should also be able to verify the authenticity of the UE. The receiver should be able to assess the validity, provenance and relevance of the data received [53], to make sure that fragmented data received is complete and not corrupted. Therefore, verifying the producer's identification to ensure that identity and source of cached data can be trusted [8] is a must.

Non-Repudiation This dimension is concerned with preventing any device from denying its involvement in an activity on the network such as denying transmitting or receiving a service [75]. It also allows the tracking of the source of a possible security violation. The SP and devices should be held accountable for their action through monitoring of network activities [49]. For example, a verified content producer should not be able to deny that they are the source of the content or UE should not be able to deny sending an interest message. Some of the solutions include the use of digital signature to achieve non-repudiation and other mechanisms should be in place to prove originality of the data as well as proof of transaction to prevent attacks such as false accusation.

Infrastructure Security Layer—*Modules 1, 2, and 3*: The non-repudiation at this layer records and identifies entities such as UE, BBU, servers that perform activities on other devices, modify control data or access UE data. This record can be used as proof of access or modification of the control data. In additional, identifies the origin of control messages and the action that was performed. Identifiers can be applied as solution to bind user related messages to the UE and network for accountability. Also, Packet Level Authentication (PLA) protocol [11] can be used to provide network layer authentication and accountability of the data using public key cryptography.

Service Security Layer—*modules 4, 5 and 6*: The non-repudiation at this layer is concerned with recording the content producer, UE, BBU or other entities that performed activities on the NS, origin of the control message and the UE that accessed the services. For example, recording information about an object's provenance, indicating the creator or publisher of content object. Moreover, the requester who receives content can be recorded and charged for the service using out-of-band digital signature solution [50]. There should be a strong association between the entity identities and use of the NS to prevent attacks like spoofing attack. However due to the nature of integrated system there might not be a direct link between SP and the requester. Therefore, the producer and requester must trust the system to account for usage in a fair manner, whereby charges are added according to services accessed periodically [9]. Traditionally, the mobile network is capable of tracking and monitoring the system utilization and usage as per the contract between the SP and the subscriber.

Data Confidentiality This dimension ensures the confidentiality of data on the UE, network devices and in transit, encryption should be used on data to provide confidentiality. Encrypted data and should only be decrypted by the authenticated and authorized entity. The confidentiality of any used data must be protected against unauthorized users or attacks such as eavesdropping and privacy invasion. Moreover, encrypting message on wireless channel will be standardized in 5G. For example, encryption keys can be applied to encrypt data using symmetric or asymmetric encryption mechanisms.

Infrastructure Security Layer—*Module 1, 2 and 3*: The data confidentiality at this layer is concerned with protecting data on the network devices and data transiting in transit from unauthorized access such as user's control and configuration data. During service provision, data might pass through possibly untrusted segments, which highlights the issues of whether the producer and requester are able to trust the infrastructure routing decisions without exposing the data. Encryption and ACL mechanisms can be used in providing data confidentiality. Other methods such as key extraction protocol based on Channel State Information (CSI) could be used to avoid leakage of data. Additional use of cryptographic mechanism like stream ciphers, might stop the attacker from reading messages between D2D users as well as preventing eavesdropping attack [34].

Service Security Layer—*Module 4, 5 and 6*: Confidentiality at this layer is concerned with protecting the NS's control, configuration and management data such as Pending Interest Table (PIT) updates, security setting from unauthorized access and modification. ACL and encryption methods can be used to provide confidentiality of NS. A content producer should be able to control which subscribers may receive what content, however confidentiality might not be relevant where the producer is offering data to everyone. Group key distribution [48] is another mechanism that can be used for data confidentiality, the producer pre-distributes keys to all potential requesters, an out-of-band approach prearrangements might be required [49].

Communication Security This dimension ensures that information only flows from source to destination endpoints using secure wireless and wired communication channels. The end user accesses resources and services by connecting to the network via wireless access point. In legacy systems, the wireless channel was not secured but in 5G this problem will be addressed [4]. However, point-to-point communication does not apply to ICN, the content is requested without being aware of its location. Also, the requester might be receiving different chunks of cached data from different sources such as content server, BBU Pool and D2D UEs which makes establishing secure connections complex and unmanageable. Therefore, information-centric and host-centric security methods must be considered, it is paramount to ensure that only intended D2D users are able to receive and read data. Encryption methods can be used to provide confidentiality, secure routing and transmission of data to authorized users. With D2D communications, physical layer security can be applied by exploiting wireless channel characteristics, modulation, coding, and multiple antennas preventing eavesdroppers [64].

Infrastructure Security Layer—*Modules 1, 2 and 3*: The communication security at this layer ensures that UE, control and management data only flows between entities and communication link that uses secure channels. For example, authentication data such as security context should not be diverted or intercepted as it flows between source and intended destination end points. Secure communication must be established between the UEs and other entities before sharing any information.

Service Security Layer—*Modules 4, 5 and 6*: The communication security at this layer ensures the management, control and UE data in transit for use by NS, only flows between entities using a secure channel and that data is not intercepted as it flows between the endpoints. For example, with the interest messages and service data, the SP registers a service identity to the server or cache node and binds the data under namespace. The BBU is tasked with monitoring and storing data regarding the interest and data exchange, BBU can search for malicious nodes and select alternatives path for packets to reach their destination securely, taking advantage of the ICN architecture which can reveal misbehaving nodes [56].

Data Integrity This dimension ensures that data is received as sent or retrieved as stored and no data manipulation has been performed by any malicious or authorized users. D2D users should be able to receives correct data without alteration or fabrication. If an attack like message injection or false reporting are initiated, the data's integrity might be violated which could compromise the UE and the whole system [45]. Data integrity can be achieved by using hash, functions, MD5, digital signature, while integrity in CCN can be provided by applying a simple content-signing method, such as the manifest-based content authentication.

Infrastructure Security Layer—*Module 1, 2 and 3*: The integrity at this layer is concerned with protecting the configuration, control data on the network entities, D2D links and data in transit or stored on the devices against unauthorized modification, creation and replication. System integrity can be compromised if an attacker uses a malicious server to insert bogus subscription and act as a bogus subscriber to the UE or BBU, then responds to interest with bogus reply or drop the data completely. Integrity is important for D2D communications to secure the user's data and enable legitimate users to decrypt the received encrypted data by using encryption.

Service Security Layer—*Module 4, 5 and 6*: The integrity at this layer is concerned with protecting the management, control, UE data against unauthorized modification, and deletion of service data. For example, the integrity of interest and service data in transits should be protected. The identifications and security context from authentication should be protected from any modification or deletion [75]. Integrity can be ensured by applying cryptographic mechanisms such as hash functions, Message Authentication Code (MAC) and data modification should be detectable, however there is no cryptographic integrity protection for the user data plane in 5G.

Availability This dimension ensures network elements and services are available to legitimate and authorized users ubiquitously. Services should be available even during attacks such as DoS and free riding [75]. In D2D communications, DoS attacks are hard to detect since the D2D does not rely on centralized infrastructure [35]. Jamming attack affects communication between D2D users and can be started anonymously [34] affecting service availability. Due to CCN naturally spreading contents to permit request being satisfied by alternating sources, it requires a lot effort to initiate a DoS attack whereby an attacker would have to send repeated requests on a single device, it is hard but possible [49]. In 5G, NS should always be available for UEs and the waiting time to connect or get services should be as short as possible to complement 5G objectives like high date rate, ultra low latency and reliability. In addition, devices such as Intrusion Detection Systems (IDS), Intrusion Prevention Systems (IPS) and firewalls should be deployed in the network as well as Business continuity (BC) and Disaster Recovery (DR) plans should in place to decrease downtime of 5G services.

Infrastructure Security Layer—*Module 1, 2 and 3*: The availability on this layer is concern with ensuring that network devices can receive control data, access UE data and manage the D2D link. Authorized UEs and other devices should have access to the infrastructure and protection against any attacks. During DoS attack, the attacker can target caching and routing plane by creating large amounts of unwanted traffic, which is cached by intermediate devices in the BBU and 5GC, resulting into cache overflow after overload of caching plane. Moreover, cache DoS attack can be initiated by polluting the content cache in CS, hence returning an incorrect content object [24]. To minimize the damage from such attack, no data should be delivered to any requests unless there is a valid subscription from the requester. Therefore, the prevention of unwanted traffic from bogus requests can improve availability and dependability whereby the system only serves valid and authorized users.

Service Security Layer—*Modules 4, 5 and 6*: The availability at this layer is concerned with ensuring that the NS are always available and accessing or managing of the NS by authorized users cannot be denied. Services must be protected from attacks such as DoS and jamming attacks. Whereby the attacker can use malicious content and subscriptions to overload the system, subscribers flood producer with bogus interest messages. Availability of services guarantees that authorized users can access the services through D2D communications. Availability and dependability can maintain satisfactory user experience.

Privacy This dimension ensures that the identifiers, user and network data is kept private. Privacy in 5G is part of a very critical security requirements, identity and location must be preserved. Privacy drives the new age of information and users data privacy has become a very sensitive topic, even though privacy in CCN architecture has not been investigated extensively but it has been in D2D communications. D2D users must be aware of which data they are sharing, and the system should collect only the required data to provide a specific service. Encryption can be used to the protect the communication and data transmission between entities.

Infrastructure Security Layer—*Module 1, 2 and 3*: The privacy at this layer is concerned with ensuring that data that can be used to identify the UE, BBU and 5GC entities or communications link is not available to unauthorized users. Network elements should not be able provide data revealing the UE's network activities such as UE's location to unauthorized user and only certain user data should be accessed by authorized personnel. Exposing information from cached data, the attacker could extract it hence violating user data privacy. The UE might need to communicate anonymously while accessing services to reduce such attacks [32].

Service Security Layer—*Module 4, 5 and 6*: The privacy at this layer is concerned with ensuring that data that can be used to identify devices, NS management systems, communication links is not available to unauthorized user. In addition, NS should not be able to reveal UE data such as UE and service identities. An attacker might be able to obtain data by monitoring cache transaction of accessed data even when the requester source is not clearly identified, this is achieved by analysing direction of the requests and timings of the transactions [49]. Location-based services can enable the tracking of the UE and data privacy might be compromised as this service relies on location of the user and service. Moreover, the UE activities can be exposed to cache owners that they might have no transactions with the UE. It is impossible for the user to request services without revealing their subscription and security information to the SP or the infrastructure. Private Information Retrieval (PIR) mechanism [71] could be used to preserve privacy of subscription data, it allows the retrieval of database entries without the user disclosing the entries to the server.

5 Security Solution Approaches

For solution approaches, we have to consider the unique characteristics of the CCN and 5G enabled D2D communications and how some of their security features are pre-designed. For instance, ICN has basic security in its architecture design such as integrity and authentication while the encryption messages over wireless communication has been standardised in 5G. In addition, the 5G trust enhancement is due to routing attacks in SS7 [25] impersonation and address spoofing attacks [61] in signalling messages which exploited the trusted domains in legacy systems. This section discusses possible security solutions to address the threats and attacks presented in this article.

5.1 Authentication and Key Management (AKM)

Authentication is a key factor in securing D2D communications, content delivery and facilitating content authenticity. A secure framework for authentication between two D2D users was proposed in [62] and [74] proposed a security communication protocol to defend systems from attacks like Man In The Middle (MITM) and masquerading. Data origin authentication method can be achieved by using of digital signature algorithm for proof of origin and protecting sensitive message from tampering. In addition, cryptography can be deployed at the different layers to achieve authentication. In this case, keys are used to encrypt and decrypt data, therefore key management plays a vital role in preservation of user and security context data. This includes the generation, distribution and storage of keys.

5.2   Confidentiality and Integrity (CI)

Data confidentiality of NS, D2D messages, control data can be implemented by using ACLs and various cryptographic techniques. While data integrity can be protected by using hash functions and digital signatures during transmission, preventing a malicious user from forging data that can affect the system's integrity. For instance, routing misuse is the result of concept of trustworthiness and integrity in CCN based on a trusted computing approach [12].

5.3   Non-repudiation Enforcement (NRE)

The use of digital signature and certificates can act as proof of work so that entities don't deny their involvement in a transaction. When the UE registers for services, the SP can monitor the services accessed and bill the subscriber accordingly. An efficient auditing system can be used to stop attack such as false accusation by logging all activities in the network with platforms such as Distributed Audit Service platforms (XDAS) [29]. The system should be able to identify the origin of false message through traceability. Additionally, the message originator can be verified through authentication process to avoid data leakage by false notifications from a malicious user.

5.4   Secure Naming, Routing, Forwarding and Transmission (SNRFT)

Content naming technique is fundamental to ICN, a verifiable binding between a content name and its provider prevents content poisoning attack. Methods such as secure naming, secure routing and forwarding techniques are vital to any network architecture security. Secure naming scheme can be achieved by using RSA and Identity Base Cryptography (IBC). A name-based method using IBC was proposed in [77] for trust management in CCN. While secure routing is essential in D2D communications especially during out of coverage, in [54] a Secure Message Delivery (SMD) protocol that protects relayed message was proposed. Whereas secure forwarding involves secure forward plane or secure namespace mapping that enable interest forwarding for name prefixes. The Interest Key Binding (IKB) method can be applied by binding the producer's public key and the content name with the interest packet [27], which maps the producer and the content.

5.5 Access Control (ACL)

Authorization enables an entity to control the access to its services requested by other entities. In mobile network, ACL can be enforced through use of RADIUS [46] and DIAMETER [13] protocols which are centralized authorization or through a distributed authorization method as proposed in [70]. Some of these ACL methods are encryption-based, attribute-based, session-based, Fine Grained Access Control (FAC) and context aware schemes. ACL can be supported by AKM to provide different levels of authorization such as access to the network and services. The authors in [72] focused on the D2D communications access, authorization was achieved by using heterogeneous and fine-grained access control mechanisms together with AKA methods. An identity-based cryptography for ACL enforcement was used in [60] while the authors in [52], proposed an ACL method for CCN based on Kerberos in IP-based networks, utilizing on distinctive authentication and authorization techniques.

5.6 Privacy Preservation (PP)

Integrating services in mobile network requires the UEs to disclose their location to the BBU for data routing and forwarding but UEs might be unwilling to share their location to avoid exposure. Privacy can be preserved by using identity expiration enforcement technique and leveraging on homomorphic cryptography [18]. The authors in [51] proposed a client anonymity framework, cryptographic based on naming scheme, it improves publisher's and consumer's privacy and untraceability. Also data privacy can be achieved through data encryption and physical layer security to define secrecy capacity as maximum transmission rate at which unintended user cannot decode transmitted message [40]. The obfuscation method is another way to preserve data privacy by degrading the quality of information like UE location to protect the user's identity.

5.7 Availability and Dependability (AD)

This approach is for both the D2D communications and services functionalities to reduce the defect attacks such as DoS and free riding that make services unavailable to legitimate users. An attacker can decrease system availability by encouraging UEs to selfishly not participate in content sharing, a cooperative mechanism between UEs was proposed in [63]. In CCN, DoS mitigation may include change in intermediate cache structure such as PIT and CS, as well as reducing the rate of consumer request through request of proof work [65]. Additionally, an interest flooding detection and mitigation method based on fuzzy logic and routers cooperation can be applied [69].

Table 3 Threats, attacks and solutions based on X.805 framework

Threats	Attacks	Solutions
Destruction of data and resources	Impersonation, MITM, routing misuse	CI, AKM
Corruption or modification of data	Content poisoning, false accusation, cache misuse, data fabrication, replay, IP/location spoofing	NRE, SNRFT
Theft, removal or loss of data resources	Unauthorized, access masquerading, false content injection, data leakage	AKM, ACL
Disclosure of data	Privacy violation, eavesdropping discovery, monitoring, timing anonymity, unlinkability, traceability	PP, AKM
Interruption of services	Cache pollution, DoS, free riding jamming, session hijacking interest flooding	AD, AKM

The presented solutions could be applied to multi layers of the network, based on the eight dimensions of X.805 framework using a modular based approach to address the vulnerabilities, threats and attacks. Also, Unconventional techniques such as cache verification and self-certifying naming methods can be applied to prevent forged content. Self-certifying is becoming a popular approach in 5G, to support network edge services, this is due its ability to handle dynamic content objects. Security on physical layer enforce security on the upper layers in D2D communications and it is necessary to study security on the high layers rather than just the physical layer. Moreover, security at upper layers is based on building security protocols to provide secure communication and data in transit without undermining D2D communications and 5G features such as network routing, caching or D2D links. Therefore, we believe the security threats in 5G require an integrated solution, hence a hybrid approach that consists of information-centric and communication-centric solutions should be used.

Many security issues in D2D communications and other NS are still open without appropriate solutions, some of the threats, attacks and possible solutions are presented in Table 3. No work is comprehensive enough to cover all security domains and fulfil all security requirements of 5G. Some of the existing work proposed solutions that achieve mobile security with IP based methods but not compatible with new use cases and services like CCN. Moreover, the security mechanisms should be lightweight to avoid high communication and computational overhead and the effects of mobility on security and privacy preservation should be considered.

6 Conclusion and Future Work

In 5G, D2D communications will be used as underlay technology to offload traffic from the backhaul to the fronthaul and push content to edge closer to the user. The secure delivery of NS to D2D users is crucial for 5G's main objectives. Due to heterogeneous nature of 5G and its enablement of new use cases, new security challenges have been created, making the secure delivery of NS difficult. Therefore, these new and old challenges need to be address using new and more robust measures. We investigated related work on the security of NS in 5G enabled D2D communications, our contribution included the introduction of NSD framework based on D2D and CCN as NS. We presented an integrated system model to investigate the security for both D2D and CCN domains, highlighting the vulnerabilities, threats, and attacks and their affect on D2D users. We then evaluated the security requirements of the system model based on X.805 security framework for a systematic and comprehensive approach. We also explored the existing approaches, then suggested a hybrid approach consisting of information-centric and host-centric solutions to provide security for hosts, the data, and the network. Most importantly, the study highlighted the lack of an integrated approach to address security for NSD in 5G enabled D2D communications which needs addressing. The open issues will motivate future research trends including security for SDN/NFV, network slicing, and integrated security solutions for 5G. Therefore, future work is to develop a comprehensive multi-layered security framework and solutions that addresses the highlighted security issues by incorporating the possible solution approaches in this article.

References

1. 3GPP (2010) Feasibility study on the security aspects of remote provisioning, change of subscription for machine to machine (m2m) equipment. Technical specification (TS) 3GPP TR 33.812 V9.2.0. Third Generation Partnership Project
2. 3GPP (2013) Feasibility study for proximity services (prose). Technical specification (TS) 3GPP TR 22.803 V12.2.0, Third Generation Partnership Project
3. 3GPP (2016) Study on architecture for next generation system. Technical specification (TS) 3GPP TR 23.799 V14.0.0 (2016-12), Third Generation Partnership Project
4. 3GPP (2020) Proximity-based services (prose); security aspects. Technical specification (TS) 3GPP TS 33.303 V16.0.0 (2020-07), Third Generation Partnership Project
5. 5GPPP (2017) Deliverable d2.7 security architecture (final). Tech. rep., 5G Enablers for Network
6. Aamir M, Zaidi SMA (2015) Denial-of-service in content centric (named data) networking: a tutorial and state-of-the-art survey. Secur Commun Netw 8(11):2037–2059. https://doi.org/10.1002/sec.1149
7. AbdAllah EG, Hassanein HS, Zulkernine M (2015) A survey of security attacks in information-centric networking. IEEE Commun Surv Tutor 17(3):1441–1454. https://doi.org/10.1109/COMST.2015.2392629
8. Ahlgren B, Dannewitz C, Imbrenda C, Kutscher D, Ohlman B (2012) A survey of information-centric networking. IEEE Commun Mag 50(7):26–36. https://doi.org/10.1109/MCOM.2012.6231276

9. Aiash M, Mapp G, Lasebae A, Loo J (2014) A secure framework for communications in heterogeneous networks. In: 2014 28th international conference on advanced information networking and applications workshops, pp 841–846. IEEE. https://doi.org/10.1109/WAINA.2014.132

10. Alliance N (2016) 5g security recommendations package 2: Network slicing. White paper

11. Andersen DG, Balakrishnan H, Feamster N, Koponen T, Moon D, Shenker S (2021) Holding the internet accountable. In: HotNets. Citeseer (2007). http://repository.cmu.edu/compsci/66. Accessed 16 January 2021

12. Anderson R (2004) Cryptography and competition policy issues with "trusted computing". Comput Secur J 20(1):1–13

13. Arkko J, Zorn G, Fajardo V, Loughney J (2021) Diameter base protocol. Rfc, IETF (2012). https://tools.ietf.org/html/rfc6733. Accessed 10 January 2021

14. Carofiglio G, Gallo M, Muscariello L, Perino D (2015) Scalable mobile backhauling via information-centric networking. In: The 21st IEEE international workshop on local and metropolitan area networks, vol 2015, pp 1–6. IEEE. https://doi.org/10.1109/LANMAN.2015.7114719

15. Chandrasekaran G, Wang N, Hassanpour M, Xu M, Tafazolli R (2018) Mobility as a service (maas): a d2d-based information centric network architecture for edge-controlled content distribution. IEEE Access 6:2110–2129. https://doi.org/10.1109/ACCESS.2017.2781736

16. Checko A, Christiansen H, Yan Y, Scolari L, Kardaras G, Berger M, Dittmann L (2015) Cloud ran for mobile networks-a technology overview. IEEE Commun Surv & Tutor 17(1):405–426. https://doi.org/10.1109/COMST.2014.2355255

17. Cisco (2017) Cisco visual networking index: Global mobile data traffic forecast update, 2016–2021 white paper. Tech rep, Cisco. https://www.cisco.com/c/en/us/solutions/collateral/service-provider/visual-networking-index-vni/mobile-white-paper-c11-520862.html. Accessed 09 September 2018

18. Dijk MV, Gentry C, Halevi S, Vaikuntanathan V, Gilbert H (2010) Fully homomorphic encryption over the integers. In: Annual international conference on the theory and applications of cryptographic techniques, vol 6110, pp 24–43. Springer Berlin Heidelberg, Berlin, Heidelberg. https://doi.org/10.1007/978-3-642-13190-5_2

19. Dolev D, Yao ACC (1983) On the security of public key protocols. IEEE Trans Inf Theory 30(2):198–208

20. Edris EKK, Aiash M, Loo J (2019) Investigating network services abstraction in 5g enabled device-to-device (d2d) communications. In: 2019 IEEE smartworld, ubiquitous intelligence computing, advanced trusted computing, scalable computing communications, cloud big data computing, internet of people and smart city innovation (SmartWorld/SCALCOM/UIC/ATC/CBDCom/IOP/SCI), pp 1660–1665. IEEE, Leicester, UK. https://doi.org/10.1109/SmartWorld-UIC-ATC-SCALCOM-IOP-SCI.2019.00296

21. Edris EKK, Aiash M, Loo J (2020) Formal verification and analysis of primary authentication based on 5g-aka protocol. In: The third international symposium on 5g emerging technologies (5GET 2020). IEEE, Paris, France

22. Edris EKK, Aiash M, Loo J (2020) Network service federated identity (ns-fid) protocol for service authorization in 5g network. In: 5th IEEE international conference on fog and mobile edge computing (FMEC 2020). IEEE, Paris, France

23. Edris EKK, Aiash M, Loo J, Alhakeem MS Formal verification of secondary authentication protocol for 5g secondary authentication. Int J Secur Netw (accepted for publication)

24. Edwall T (2019) The network of information: architecture and applications. Tech Rep, SAIL Project Team. https://sail-project.eu/wp-content/uploads/2011/08/SAIL_DB1_v1_0_final-Public.pdf. Accessed 02 June 2019

25. Engel T (2014) Ss7: Locate. track. manipulate. In: Talk at 31st chaos communication congress

26. Ghali C, Tsudik G, Uzun E (2014) Elements of trust in named-data networking. ACM SIGCOMM Comput Commun Rev 44:12–19. https://doi.org/10.1145/2677046.2677049

27. Ghali C, Tsudik G, Uzun E (2014) Needle in a haystack: mitigating content poisoning in named-data networking. In: NDSS symposium. https://doi.org/10.14722/sent.2014.23014

28. Golrezaei N, Molisch AF, Dimakis AG, Caire G (2013) Femtocaching and device-to-device collaboration: A new architecture for wireless video distribution. Commun Mag IEEE 51(4):142–149. https://doi.org/10.1109/MCOM.2013.6495773
29. Group O (1998) Distributed audit service (xdas) (1998). http://www.opengroup.org/security/das/xdas_int.htm. Accessed 05 June 2019
30. Guey JC, Liao PK, Chen YS, Hsu A, Hwang CH, Lin G (2015) On 5g radio access architecture and technology [industry perspectives]. Wirel Commun. IEEE 22(5):2–5. https://doi.org/10.1109/MWC.2015.7306369
31. Gupta A, Jha RK (2015) A survey of 5g network: architecture and emerging technologies. IEEE Access 3:1206–1232. https://doi.org/10.1109/ACCESS.2015.2461602
32. Hamoud ON, Kenaza T, Challal Y (2018) Security in device-to-device communications: a survey. IET Netw 7(1):14–22. https://doi.org/10.1049/iet-net.2017.0119
33. Han B, Hui P, Kumar VA, Marathe MV, Pei G, Srinivasan A (2010) Cellular traffic offloading through opportunistic communications: a case study. In: Proceedings of the 5th ACM workshop on challenged networks, pp 31–38. https://doi.org/10.1145/1859934.1859943
34. Haus M, Waqas M, Ding AY, Li Y, Tarkoma S, Ott J (2017) Security and privacy in device-to-device (d2d) communication: a review. IEEE Commun Surv & Tutor 19(2):1054–1079. https://doi.org/10.1109/COMST.2017.2649687
35. Huang H, Ahmed N, Karthik P (2011) On a new type of denial of service attack in wireless networks: the distributed jammer network. IEEE Trans Wirel Commun 10(7):2316–2324. https://doi.org/10.1109/TWC.2011.052311.101613
36. Jacobson V (2009) A description of content-centric networking (ccn). Future Internet Summer School (FISS). https://named-data.net/publications/van-ccn-bremen-description/. Accessed 10 March 2021
37. Jin H, Xu D, Zhao C, Liang D (2017) Information-centric mobile caching network frameworks and caching optimization: a survey. EURASIP J Wirel Commun Netw 2017(1):1–32. https://doi.org/10.1186/s13638-017-0806-6
38. Kang HJ, Kang CG (2014) Mobile device-to-device (d2d) content delivery networking: a design and optimization framework. J Commun Netw 16(5):568–577. https://doi.org/10.1109/JCN.2014.000095
39. Lee YL, Loo J, Chuah TC, Wang LC (2018) Dynamic network slicing for multitenant heterogeneous cloud radio access networks. IEEE Trans Wirel Commun 17(4):2146–2161. https://doi.org/10.1109/TWC.2017.2789294
40. Leung-Yan-Cheong S, Hellman ME (1978) The gaussian wire-tap channel. IEEE Trans Inf Theory 24(4):451–456. https://doi.org/10.1109/TIT.1978.1055917
41. Liang C (2015) Wireless virtualization for next generation mobile cellular networks. IEEE Wirel Commun Mag 22(1):61–69. https://doi.org/10.1109/MWC.2015.7054720
42. Liang C, Yu FR, Zhang X (2015) Information-centric network function virtualization over 5g mobile wireless networks. Network. IEEE 29(3):68–74. https://doi.org/10.1109/MNET.2015.7113228
43. Liang C, Yu F (2015) Wireless network virtualization: a survey, some research issues and challenges. IEEE Commun Surv Tutor 17(1):358–380. https://doi.org/10.1109/COMST.2014.2352118
44. Lichtman M, Rao R, Marojevic V, Reed J, Jover RP (2018) 5g nr jamming, spoofing, and sniffing: Threat assessment and mitigation. In: 2018 IEEE international conference on communications workshops (ICC Workshops), pp 1–6. IEEE
45. Lin X (2009) Cat: building couples to early detect node compromise attack in wireless sensor networks. In: GLOBECOM 2009-2009 IEEE global telecommunications conference, pp 1–6. IEEE (2009). https://doi.org/10.1109/GLOCOM.2009.5425922
46. Lior A, DeKok A (2013) Remote authentication dial in user service (radius) protocol extensions. Rfc, IETF.https://tools.ietf.org/html/rfc6929. Accessed 05 February 2019
47. Liu D, Chen B, Yang C, Molisch AF (2016) Caching at the wireless edge: design aspects, challenges, and future directions. Commun Mag. IEEE 54(9):22–28. https://doi.org/10.1109/MCOM.2016.7565183

48. Liu H, Chen Z, Tian X, Wang X, Tao M (2014) On content-centric wireless delivery networks. IEEE Wirel Commun Mag 21(6):118–125. https://doi.org/10.1109/MWC.2014.7000979
49. Loo J, Aiash M (2015) Challenges and solutions for secure information centric networks: a case study of the netinf architecture. J Netw Compur Appl 50:64–72. https://doi.org/10.1016/j.jnca.2014.06.003
50. Mao W (2004) Modern cryptography: theory and practice. Prentice Hall PTR, Upper Saddle River, N.J
51. Martinez-Julia P, Gomez-Skarmeta A (2012) Using identities to achieve enhanced privacy in future content delivery networks. Comput Electr Eng 38(2):346–355. https://doi.org/10.1016/j.compeleceng.2011.11.021
52. Nunes IO, Tsudik G (2018) Krb-ccn: lightweight authentication & access control for private content-centric networks. In: International conference on applied cryptography and network security, pp 598–15. Springer (2018)
53. Ohlman B, Davies E, Spirou S, Pentikousis K, Boggia G (2014) Information-centric networking: evaluation methodology. IETF (The Internet Engineering Task Force) request for comments. https://tools.ietf.org/html/draft-irtf-icnrg-evaluation-methodology-01. Accessed 05 January 2021
54. Panaousis E, Alpcan T, Fereidooni H, Conti M (2014) Secure message delivery games for device-to-device communications. In: Poovendran R, Saad W (eds) Decision and game theory for security, pp 195–215. Springer International Publishing, Switzerland (2014). https://doi.org/10.1007/978-3-319-12601-2_11
55. Park Y, Park T (2007) A survey of security threats on 4g networks. In: IEEE globecom workshops, pp 1–6. IEEE. https://doi.org/10.1109/GLOCOMW.2007.4437813
56. Priya V, Sakthisaravanan B (2015) Information centric network for secure data transmission in dtn. In: International conference on innovation information in computing technologies, pp 1–4. IEEE. https://doi.org/10.1109/ICIICT.2015.7396101
57. Raheem A, Lasebae A, Aiash M, Loo J (2013) Supporting communications in the iots using the location/id split protocol: a security analysis. In: Second international conference on future generation communication technologies (FGCT 2013), pp. 143–147. IEEE
58. Ravindran R (2019) Enabling icn in 3gpp's 5g nextgen core architecture. IETF (The Internet Engineering Task Force) Request for Comments. https://tools.ietf.org/id/draft-ravi-icnrg-5gc-icn-00.html. Accessed 03 March 2021
59. Ravindran R, Chakraborti A, Amin SO, Azgin A, Wang G (2017) 5g-icn: Delivering icn services over 5g using network slicing. IEEE Commun Mag 55(5):101–107. https://doi.org/10.1109/MCOM.2017.1600938
60. Raykova M, Lakhani H, Kazmi H, Gehani A (2015) Decentralized authorization and privacy-enhanced routing for information-centric networks. In: Proceedings of the 31st annual computer security applications conference, vol 7–11, pp 31–40. https://doi.org/10.1145/2818000.2818001
61. RIFS G (2016) Diameter roaming security—proposed permanent reference document. Tech rep, GSMA (2016)
62. Shen W, Hong W, Cao X, Yin B, Shila DM, Cheng Y (2014) Secure key establishment for device-to-device communications. IEEE Global Commun Conf. https://doi.org/10.1109/GLOCOM.2014.7036830
63. Sun J, Chen X, Zhang J, Zhang Y, Zhang J (2014) Synergy: a game-theoretical approach for cooperative key generation in wireless networks. In: IEEE INFOCOM 2014-IEEE conference on computer communications, pp 997–1005. IEEE. https://doi.org/10.1109/INFOCOM.2014.6848029
64. Sun L, Du Q (2017) Physical layer security with its applications in 5g networks: a review. Communications, China 14(12):1–14. https://doi.org/10.1109/CC.2017.8246328
65. Tourani R, Misra S, Mick T, Panwar G (2018) Security, privacy, and access control in information-centric networking: a survey. IEEE Commun Surv Tutor 20(1):566–600. https://doi.org/10.1109/COMST.2017.2749508

66. Tran TX, Hajisami A, Pompili D (2017) Cooperative hierarchical caching in 5g cloud radio access networks. IEEE Netw 31(4):35–41. https://doi.org/10.1109/MNET.2017.1600307
67. Wang K, Yu FR, Li H, Li Z (2017) Information-centric wireless networks with virtualization and d2d communications. IEEE Wirel Commun 24(3):104–111. https://doi.org/10.1109/MWC.2017.1500384WC
68. Wang M, Yan Z, Niemi V (2017) Uaka-d2d: Universal authentication and key agreement protocol in d2d communications. Mob Netw Appl 22(3):510. https://doi.org/10.1007/s11036-017-0870-5
69. Wang Y, Xu M, Feng Z, Li Q, Li Q (2014) Session-based access control in information-centric networks: Design and analyses. In: 2014 IEEE 33rd international performance computing and communications conference (IPCCC), pp 1–8. IEEE. https://doi.org/10.1109/PCCC.2014.7017094
70. Woo TY, Lam SS (1998) Designing a distributed authorization service. In: Proceedings. IEEE INFOCOM'98, the conference on computer communications. seventeenth annual joint conference of the ieee computer and communications societies. Gateway to the 21st Century (Cat. No. 98), vol 2, pp 419–429. IEEE
71. Yi X, Paulet R, Bertino E (2013) Private information retrieval. Synth Lect Inform Secur Priv Trust 4(2):1–114. https://doi.org/10.2200/S00524ED1V01Y201307SPT005
72. Yue J, Ma C, Yu H, Zhou W (2013) Secrecy-based access control for device-to-device communication underlaying cellular networks. Commun Lett IEEE 17(11):2068–2071. https://doi.org/10.1109/LCOMM.2013.092813.131367
73. Zeltsan Z (2005) Security architecture for systems providing end-to-end communications. http://www.itu.int/ITU-T/worksem/ngn/200505/presentations/s5-zeltsan.pdf. Accessed 05 January 2019
74. Zhang A, Chen J, Hu RQ, Qian Y (2016) Seds: secure data sharing strategy for d2d communication in lTE-advanced networks. IEEE Trans Veh Technol 65(4):2659–2672. https://doi.org/10.1109/TVT.2015.2416002
75. Zhang T, Fan H, Loo J, Liu D (2017) User preference aware caching deployment for device-to-device caching networks. IEEE Syst J 13(1):226–237
76. Zhang T, Fang X, Liu Y, Nallanathan A (2019) Content-centric mobile edge caching. IEEE. Access 8:11722–11731
77. Zhang X, Chang K, Xiong H, Wen Y, Shi G, Wang G (2011) Towards name-based trust and security for content-centric network. In: 2011 19th IEEE international conference on network protocols, pp 1–6. IEEE. https://doi.org/10.1109/ICNP.2011.6089053

Digital Leadership, Strategies and Governance in Cyberspace

Jim Seaman

Abstract For centuries, Command and Control is a model that has been success-fully employed by many countries' military forces, in defense of their valuable military assets that support their various missions—both at home and on overseas deployments. Businesses can gain considerable benefits through the adoption of a Command and Control style structure throughout their organization and in this chapter the description of a Command and Control structure and the potential benefits to protecting a digital company's Cyberspace operations are described.

Keywords Leadership · Management · Asset management · Risk management · Military doctrine · Teamwork · Policies · Procedures · Security · Defense

1 Introduction

One of the most important aspects of cyber-security is the way the 'Tone at the Top' is established. Do the executive board members show an interest in cyber-security and do they lead by example when it comes to acting in a manner for which the company expects—in line with the company's policies and procedures?

Without effective leadership, strategy, and governance being established, it can often feel like cyber-security is a 'tick-box' exercise or something that the business is being pressed into doing.

For the cyber-security manager, this can feel like they are pushing a heavy boulder up a very steep hill. Imagine what it must be like to work in an organization that states that it wants to be secure but then the senior managers or business executives undermine the security measures because it is not convenient for them.

For example,

- Changing the rules around password management, so that they need not change their passwords or can reuse the same passwords.

J. Seaman (✉)
IS Centurion Consulting Ltd., Castleford, UK
e-mail: contact@iscenturion.com

© The Author(s), under exclusive license to Springer Nature Switzerland AG 2021
R. Montasari et al. (eds.), *Challenges in the IoT and Smart Environments*,
Advanced Sciences and Technologies for Security Applications,
https://doi.org/10.1007/978-3-030-87166-6_2

- Not being required to book in their visitors.
- Being able to use their devices for business use when the policy prohibits this.

Consequently, it is important to understand these terms and how they relate to the protection of cyberspace.

2 Understanding the Terminology

2.1 Cyberspace [1]

"The online world of computer networks and especially the Internet, the environment in which communication over computer networks occurs," 1982, often written as two words at first, coined by science fiction writer William Gibson (best known for "Neuromancer") and used by him in a short story published in 1982, from cyber- + space (n.).

By applying the true nature of this term, another description of cyberspace could be associated with the term 'Internet of Things' [2]:

The networking capability that allows information to be sent to and received from objects and devices (such as fixtures and kitchen appliances) using the Internet.

Consequently, the term cyberspace is an extremely broad term and relates to any piece of technology which is connected to and communicating across the internet.

2.2 Digital [3]

mid-15c., "pertaining to numbers below ten;" 1650s, "pertaining to fingers," from Latin digitalis, from digitus "finger or toe" (see digit). The numerical sense is because numerals under 10 were counted on fingers. Meaning "using numerical digits" is from 1938, especially of computers which run on data in the form of digits (opposed to analogue) after c. 1945. In reference to recording or broadcasting, from 1960.

Taken from its origins, the term Digital has changed significantly since the advancements of technology and it has now drawn differing interpretations as to what it means for the individuals—based upon the perspectives [4]:

For some executives, it's about technology.
For others, digital is a new way of engaging with customers.
And for others still, it represents an entirely new way of doing business.

As you can see from these differences, it is extremely important to understand and appreciate what the term digital means for your audience, to ensure that this is being interpreted in the same manner, so that any approaches are aligned.

2.3 Leadership [5]

1821, "position of a leader, command," from leader + -ship. Sense extended by late 19c. to "characteristics necessary to be a leader, capacity to lead."

The Royal Air Force (RAF) recognize the importance of having good and effective leadership and continually seek to identify natural leadership skillsets and to reinforce its importance, especially in a fast-paced, changing world [6].

In a world that is now changing faster than ever, where technology is advancing rapidly, the RAF needs leaders who are flexible in approach and able to consider new ways of doing things.

RAF leaders must be open minded, responsive to change, constantly looking for the opportunities that change brings and be able to cope with the discomfort that is associated with change.

It is important to understand the differences between good leadership and good management. For example, do you understand the different traits of leadership versus management? [7].

Four important traits of a manager:

1. **The ability to execute a Vision**
   Managers build a strategic vision and break it down into a roadmap for their team to follow.
2. **The ability to Direct**
   Managers are responsible for day-to-day efforts while reviewing necessary resources and anticipating needs to make changes along the way.
3. **Process Management**
   Managers have the authority to establish work rules, processes, standards, and operating procedures.
4. **People Focused**
   Managers are known to look after and cater to the needs of the people they are responsible for: listening to them, involving them in certain key decisions, and accommodating reasonable requests for change to contribute to increased productivity.

Five important traits of a leader:

1. **Vision**
   A leader knows where they stand, where they want to go and tend to involve the team in charting a future path and direction.
2. **Honesty and Integrity**
   Leaders have people who believe them and walk by their side down the path the leader sets.
3. **Inspiration**
   Leaders are usually inspirational—and help their team understand their own roles in a bigger context.

4. *Communication Skills*
   Leaders always keep their team informed about what is happening, both present and the future—along with any obstacles that stand in their way.
5. *Ability to Challenge*
   Leaders are those that challenge the status quo. They have their style of doing things and problem-solving and are usually the ones who think outside the box.

2.4   Strategy [8]

1810, "art of a general," from French stratégie (18c.) and directly from Greek strategia "office or command of a general," from strategos "general, commander of an army," also the title of various civil officials and magistrates, from stratos "multitude, army, expedition, encamped army," literally "that which is spread out" (from PIE root *stere- "to spread") + agos "leader," from agein "to lead" (from PIE root *ag- "to drive, draw out or forth, move"). In non-military use from 1887.

When looking to develop your strategy, this needs to be flexible and adaptable, whilst representing a collaborative effort, which represents an alignment to the interests of the business key stakeholders. This will help to ensure that the strategy has the key stakeholder buy in and ensures that it represents a joined-up approach to the mitigation of existing, and newly identified, threats against key business interests.

2.5   Governance [9]

Late 14c., "act or manner of governing," from Old French governance "government, rule, administration; (rule of) conduct" (Modern French gouvernance), from governer "to govern, rule, command" (see govern). Fowler writes that the word "has now the dignity of incipient archaism," but it might continue useful in its original sense as government comes to mean primarily "the governing power in a state."

This is often the area in which most organizations have 'Governance' in paper *(or say that they have 'Good Governance')* but, in practice, fail to formally embed 'Governance' against the safeguarding of valuable business assets or against sensitive data operations. Consequently, this becomes a defensive program of works that is reliant upon a handful of 'trusted' personnel.

However, what happens if this small pool of employees is not in the right place, at the right time, or are unavailable at the very time that they are needed?

Consequently, a better approach is to adopt something like the military's highly successful 'Command and Control' defensive structure [10]:

To accomplish their missions, military leaders must be able to command and control the many activities of their forces.

RAND has applied strategic analysis since its earliest work on game theory to develop scenarios and guide military and civilian decisionmakers on the most effective employment of command and control (C2) principles and technologies, and continues to do so with C4I (command, control, communications, computers, and intelligence).

3 Command and Control

The benefits of an effective Command and Control (C2) structure can be seen in use many centuries ago, when used by the Roman Empire and has been actively used throughout many military establishments, for the proactive defense of valued assets, which provide critical support to military operations and missions.

In fact, Arrian himself, recognized the need for personal, hands-on leadership [11]:

> *The commander of the entire army, Xenophon [Le. Arrian], should lead from a position well in front of the infantry standards;*
>
> *He should visit all the ranks and examine how they have been drawn up.*
>
> *He should bring order to those who are in disarray and praise those who are properly drawn up.*

Just what is C2, I hear you ask. The Oxford University Press explain the term as follows [12]:

> *In the military, the term command and control (C2) means a process (not the systems, as often thought) that commanders, including command organizations, use to plan, direct, coordinate, and control their own and friendly forces and assets to ensure mission accomplishment.*
>
> *Command and control of U.S. armed forces today is the result of a long historical evolution. From 1775 to 1947, there was no common superior to the War and Navy Departments and their respective military services, except for the president of the United States, who was commander in chief of the army and navy under the Constitution. Only in 1947 were all the military departments and services unified in principle.*

As you can see, this is not just a case of having a suite of security tools at a business' disposal, but rather an organized team of individuals with the right level of skills and experiences, along with effective security toolsets, who are able to work together in the defense of the organization—ensuring the continuity of its valued business operations, in support of the company mission, objectives or values.

If the Romans saw the value of an established and effective C2 system many centuries ago and the military are still employing this type of model today, doesn't it make sense for modern business to do likewise, in defense of their empires?

Let's face it, considering the increasingly digital and data dependent businesses, effective C2 has never been more important, and an effective defensive structure should not be limited to a handful of personnel, and these businesses need to develop a holistic structure that delegates security responsibilities for everyone in the business. However, the delegation of responsibilities should be tiered based upon the potential risks that are associated against their individual job roles.

For example, everyone with access to business IT systems, email and the internet should understand, and be aware of, their responsibilities to protect their frontend access (e.g. Secure Password Management, Control of Access Credentials, etc.), be alert to malicious emails and to avoid visiting malicious websites or downloading malicious software.

3.1 C2 Definition

Global Security defines C2 as being [13]:

> Command and control is the exercise of authority and direction by a properly designated commander over assigned and attached forces in the accomplishment of a mission.
>
> Commanders perform command and control functions through a command and control system. This definition leads to several conclusions (see Fig. 1):

- The focus of C2 is the commander. Commanders assess the situation, make decisions, and direct actions.
- The goal of C2 is mission accomplishment. The main criterion of success for C2 is how it contributes to achieving that goal. Other criteria may include positioning the force for future operations and using resources effectively.

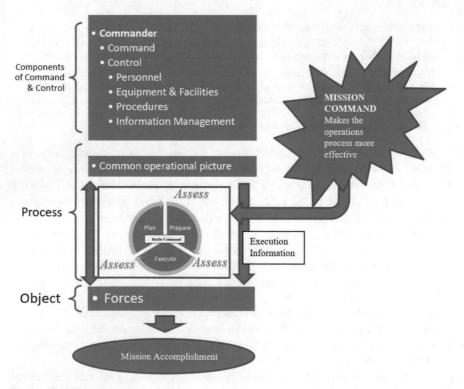

Fig. 1 C2 structure

- *C2 is directed toward forces-combat, combat support, and combat service support. Said another way, forces are the object of C2.*
- *Commanders exercise authority and direction over forces by establishing command or support relationships.*
- *Commanders must dedicate and organize resources for exercising C2. Commanders use these resources to plan and continuously assess operations that the force prepares for and executes.*
- *The commander's C2 system manages information to produce and disseminate a common operational picture (COP) to the commander, team, and subordinate forces.*
- *The C2 system supports the commander in directing forces by transmitting execution information.*

Your C2 structure involves the formal arrangement of the supporting personnel, information management, procedures, equipment and facilities that are needed for the Commander (Board Members) to conduct secure and resilient business operations.

Your C2 structure supports the Commander by performing the following three functions, in support of secure and resilient operations:

1. Creating and maintaining the common operational picture (COP).
2. Supporting decision making by improving its speed and accuracy.
3. Supporting preparation and communication of execution information.

4 Components of a C2 System

The function of an effective CR systems is to ensure that the defense of your organization's cyberspace presence employs a collaborative approach. Consequently, this does not rely on a 'Lone Wolf' to help safeguard the continual management of safe systems and processes.

Whether your employees work in the 'Engine Room' or 'Above Deck', they all have some responsibilities for ensuring the safety and security of your business. Thus, the C2 System combines elements from the Personnel; Information Management; Procedures and the Equipment and Facilities, as shown at Table 1.

Table 1 C2 System elements

C2 System				
Personnel	Information management		Procedures	Equipment and facilities
	Relevant information	Information security		
		IT and network systems	Communications	

As you can see an effective C2 system involves an integrated command traversing from higher, lower and lateral business perspectives, providing an integrated relationship of the information management (IM) activities:

- Collection.
- Display.
- Processing.
- Storing.
- Transmission.

Appropriately locating the C2-system elements is important to the effectiveness of the C2-system. Reliable inter-department communications-together with administrative support to the commander and team-are vital to C2-system continuity and effectiveness.

The C2 facilities are high-value targets for enemies, their security is important. Commanders consider the following characteristics when placing the physical C2-system elements:

- **Communications**.
  Departments should offer good communications to higher, lower, supporting, supported, and adjacent business operations.
- **Security**.
  C2 facilities must provide security for personnel and equipment.
- **Concealment**.
  Effective concealment contributes to security. Your most valuable assets should not be unprotected, when in the public domain.
- **Accessibility**.
  Assets should be only accessible by authorized personnel and personnel that are granted such access privileges should understand the need to safeguard this access.
      The design of the C2 system needs to be proportionate to the perceived risks of the assets that support the business mission statement. Consequently, the following areas should be considered in the design and development of a suitable C2 system.
- **Flexibility**.
  The number of Information Systems available and their size, weight, and power considerations all affect the flexibility. C2-system flexibility must match that of the business interests.
- **Continuity of command**.
  The C2 system must function 24/7, all year round to ensure that timely identification and reaction to impactful incidents caused by new or existing threats.
- **Fusion of command and team effort**.
  An effective C2 system integrates and facilitates command and team efforts.
      The supporting assets, as well as the procedures, should facilitate an integrated approach, ensuring lateral communication among various departments and vertical communication between them and the Commander.

- **Size**.
  The Commander should balance flexibility and longevity when deciding the size of the C2 system. A larger C2 system may provide greater flexibility and longevity through greater redundancy capabilities. However, this comes at the cost of potentially slower decision-making, greater resource investment, and decreased agility, security, flexibility, and mobility.

  A smaller C2 system may limit the capability but increase survivability and mobility. Several smaller dispersed facilities may provide equal redundancy and greater survivability than one large facility.

  The key is to strike the right balance and provide a responsive yet agile organization.

  Commanders identify necessary elements and eliminate unnecessary ones.
- **Resilience**.
  Resilience refers to the degree of protective measures provided by the C2 system, primarily by facilities and equipment. Resilience extends beyond placing valued assets and sensitive data behind effective protective measures; it involves a combination of active and passive measures, with operations security (OpSec) measures being essential to both the entire C2 system and the individual business assets that it seeks to defend.
- **Modularity**.
  Incorporating modularity in the C2-system design offers flexibility in the deployment and employment of the C2 system. The Commander should tailor the C2 system to the business mission.

  Commanders should add or remove elements, as required to provide continued support. However, when separating C2-system elements, Commanders should balance the advantages of remote-working practices against the disadvantages of loss of personal contact and face-to-face planning.
- **Capacity**.
  A C2 system requires enough Information Management capacity to manage the Relevant Information the business needs to continue operating effectively.

  Information Management includes timely passage of Relevant Information to all who need it.
- **Survivability**.
  A C2 system must be reliable, robust, and resilient. It must be at least as survivable as the business itself. Distributed systems and back-up systems meet these requirements.

  The Commander should organize and deploy the C2 system so that performance under stress degrades gradually, not catastrophically.

  The C2 system must be able to cope with systems degradation or failure.
- **Geography**.
  The C2 system requires Information Systems with enough range to link all departments with which the commander communicates, including those outside the business' area of operations (AO).

  Increasingly, this means providing a reach back capability to business headquarters.

This may require dispersed systems.

- **Mobility**.

In a modern digital business, connected to Cyberspace, The C2 system needs to be as mobile as the overall business operations. Some C2-system elements, especially those that provide dispersed operations and connectivity to the rest of the for other parts of the business may need to move more quickly.

- **Interoperability**.

For unified actions, Information Systems need to be compatible and interoperable, especially where bring your own device (BYOD) or use your own device (UYOD) initiatives are in place.

Business Information Systems need to work with employee-owned information systems/devices, particularly as part of contingency operations.

During these operations, business and employee-owned Information Systems might be integrated, for example, with third party Information Systems.

- **Longevity**.

An effective C2 system integrates and facilitates the close coordination between the Commander and asset life-cycle planners, so that the Commander understands the risks associated with valued systems that are approaching their end of life.

Additionally, advances in technology enhance the ability for high-performance business operations, by ensuring that key supporting assets provide for maximal operational reach and sustainment.

4.1 Personnel

Your employees and customers represent the most important component of an effective C2 system, as these act as the 'eyes and ears' of the organization and are frequently the target for social engineering style attacks.

They assist you Commanders and their adherence to the company rules helps to reduce the risks to the business and they will provide vital support the various security steering and risk committee members, deputy commanders (department security champions/leads), second in command (deputy department security champions/leads), as well as to the various specialist security support roles, who provide essential support to business departments and Commanders.

An effective C2 system accounts for the characteristics and limits of human nature to create effective teams, to utilize develop the uniquely human skills and abilities.

People have individual characteristics and skills that need to be identified and honed to build effective team, capable of providing resilient capabilities able to effectively respond to and manage impactful events.

In developing an effective team, understanding their personality types can be extremely beneficial to the success or failure of these teams, helping to linking together the compatible cogs.

Table 2 Belbin team roles

Resource investigator	*Uses their inquisitive nature to find ideas to bring back to the team*
	Strengths: *Outgoing, enthusiastic. Explores opportunities and develops contacts* **Allowable weaknesses**: *Might be over-optimistic, and can lose interest once the initial enthusiasm has passed* **Don't be surprised to find that**: *They might forget to follow up on a lead*
Team worker	*Helps the team to gel, using their versatility to identify the work required and complete it on behalf of the team* **Strengths**: *Co-operative, perceptive and diplomatic. Listens and averts friction* **Allowable weaknesses**: *Can be indecisive in crunch situations and tends to avoid confrontation* **Don't be surprised to find that**: *They might be hesitant to make unpopular decisions*
Co-ordinator	*Needed to focus on the team's objectives, draw out team members and delegate work appropriately* **Strengths**: *Mature, confident, identifies talent. Clarifies goals* **Allowable weaknesses**: *Can be seen as manipulative and might offload their own share of the work* **Don't be surprised to find that**: *They might over-delegate, leaving themselves little work to do*
Plant	*Tends to be highly creative and good at solving problems in unconventional ways* **Strengths**: *Creative, imaginative, free-thinking, generates ideas and solves difficult problems* **Allowable weaknesses**: *Might ignore incidentals, and may be too preoccupied to communicate effectively* **Don't be surprised to find that**: *They could be absentminded or forgetful*
Monitor evaluator	*Provides a logical eye, making impartial judgements where required and weighs up the team's options in a dispassionate way* **Strengths**: *Sober, strategic and discerning. Sees all options and judges accurately* **Allowable weaknesses**: *Sometimes lacks the drive and ability to inspire others and can be overly critical* **Don't be surprised to find that**: *They could be slow to come to decisions*
Specialist	*Brings in-depth knowledge of a key area to the team* **Strengths**: *Single-minded, self-starting and dedicated. They provide specialist knowledge and skills* **Allowable weaknesses**: *Tends to contribute on a narrow front and can dwell on the technicalities* **Don't be surprised to find that**: *They overload you with information*

(continued)

Table 2 (continued)

Shaper	Provides the necessary drive to ensure that the team keeps moving and does not lose focus or momentum **Strengths**: Challenging, dynamic, thrives on pressure. Has the drive and courage to overcome obstacles **Allowable weaknesses**: Can be prone to provocation, and may sometimes offend people's feelings **Don't be surprised to find that**: They could risk becoming aggressive and bad-humored in their attempts to get things done
Implementer	Needed to plan a workable strategy and carry it out as efficiently as possible **Strengths**: Practical, reliable, efficient. Turns ideas into actions and organizes work that needs to be done **Allowable weaknesses**: Can be a bit inflexible and slow to respond to new possibilities **Don't be surprised to find that**: They might be slow to relinquish their plans in favor of positive changes
Completer finisher	Most effectively used at the end of tasks to polish and scrutinize the work for errors, subjecting it to the highest standards of quality control **Strengths**: Painstaking, conscientious, anxious. Searches out errors. Polishes and perfects **Allowable weaknesses**: Can be inclined to worry unduly, and reluctant to delegate **Don't be surprised to find that**: They could be accused of taking their perfectionism to extremes

For example, by carrying out a psychometric test (such as Belbin) can help you to identify the team members against the nine team roles, as depicted in Table 2 (Belbin Associates 2021).

Defending your organization requires a team effort and, whether your part of the frontend or backend business operations, the defense of the business will involve you.

Consequently, it is essential that all personnel are considered for inclusion in your business' efforts to defend your company navigate the perils of cyberspace.

4.2   Security and Risk Steering Committees

Your security and risk steering committee members are extremely important for supporting your Commanders and in helping to make and implement decisions that are aligned with the interests of the company and which reduce the risk to with the tolerable levels for the Commanders (Risk Owners).

These committees should be diverse enough to provide sufficient representation from across the business and should not be limited to just the representation of the IT and Security teams.

The committee members provide relevant information and analysis, make estimates and recommendations, prepare plans and orders and monitor execution.

The Commanders provide their security and risk committee with leadership, direction, and guidance and each committee member undertakes all their activities on behalf of the commander. However, the committees have no authority by themselves but derive authority from the commander and exercises this only in the commander's name. Commanders use their committees to exercise C2 when they cannot do so personally.

With larger committees, it can take longer for them to perform their functions and, as a result, it becomes more important to ensure that any committee meetings are structured and constructive.

Roles and Responsibilities

Each employee and committee members need to understand the importance of their role, in defense of cyberspace, which should be formally documented and with each individual having read and understood their role and responsibilities.

Your steering committees are responsible for the effective operation of the commander's C2 system and all committee members, employees and the supporting procedures exist to fulfill the following three functions:

1. Support the commander.

A committee member's most important function is to support and advise the commander throughout the operations process. It does this through information management, which includes each representative section providing control over its field of interest.

Commanders structure formal committee processes to provide the two types of information associated with decision making:

- Common operational picture-related information.
- Execution information.

All other committee member activities are secondary to their primary job roles.

A key product of the steering committees is to cascade out safe business practices and to act as a conduit for enhanced security culture, responsible leading situational understanding, collecting data and extracting any relevant information to give commanders only what they need to achieve and maintain situational understanding and make decisions.

All these efforts help commanders identify critical requirements and achieve accurate situational understanding so that they can initiate protective activities faster than threat actors are able to exploit any weaknesses.

The members of the steering committees should also prepare and disseminate execution information on behalf of the commanders, to help to communicate most of this information in the form of plans and orders.

Committee members shall communicate the commander's decisions, and provide the intent behind them, efficiently and effectively throughout the organization to help keep the business focused on accomplishing the mission.

Finally, each committee representative provides control over their field of interest throughout the operations process. While commanders make the key decisions, they are not the only decision makers. Trained, trusted committee members, given authority for decisions, and execution based on the commander's intent, free commanders from routine decisions to focus on key aspects of the operation.

This practice furthers mission command. Standing operating procedures (SOPs) may establish these responsibilities, or commanders may delegate them for specific situations.

2. Assist subordinate units.

While the committee member's priority is assisting the commander, it also assists subordinate units. An effective committee enhances subordinate units' ability to operate securely. A proficient committee works in an effective, efficient, and cooperative manner with other department representatives. It assists subordinate departments by providing resources the commander allocates to them, representing subordinates' concerns to the commander, clarifying orders and directives, and passing relevant information quickly.

Effective committees establish and maintain a high degree of coordination and cooperation with other committee members of higher, lower, supporting, supported, and adjacent business departments. This relationship is based on mutual respect, developed through a conscientious, determined, and helpful approach focused on solving problems. Anything less undermines the confidence and trust required for mission command at all levels. Favorable personal interactions among all members of a committee, and with the other committee representatives, cultivate the desired relationship.

3. Keep subordinate, higher, adjacent, supported, and supporting headquarters informed.

Committee members pass all relevant information to other departments as soon after determining the information's value to the recipient as possible. The key is relevance, not volume. Masses of data of data are worse than meaningless; they inhibit C2 by distracting team members from relevant information.

Effective information management identifies the information the commander and each committee representative require, and its relative importance. Information should reach recipients based on their need for it. Sending incomplete information sooner is better than sending complete information too late. When forwarding information, senders highlight key information for each recipient and clarify the commander's intent. Senders may pass information directly, include their own analysis, or add context to it. Common, distributed databases can accelerate this function; however, they cannot replace the personal contact that adds context.

Keeping other departments informed contributes to situational understanding at all departments. While commanders are responsible for keeping their higher and subordinate commanders informed, committee members supplement their commanders' direct communications by providing clarification through team and technical channels.

All other information goes through team or technical channels. When authorized, committee members may also inform their counterparts at other business departments of information being passed between commanders. This helps the key stakeholders better support its commander.

A Team Effort

Commander and Business Departments.

Commanders are responsible for all their business departments do (or fail to do). A commander is not able to delegate this responsibility. The final decision, as well as the final responsibility, remains with the commander (They remain accountable). When commanders assign a department member a task, they delegate the authority necessary to accomplish it. Commanders still provide guidance, resources, and support. They help to foster an organizational climate of mutual trust, cooperation, and teamwork.

Deciding and acting faster than the threat actors requires commanders and departments to focus on anticipating and recognizing the threat actors' activities (Threat Intelligence). Although commanders set the pace as the principal decision makers, their relationship with their departments must be one of loyalty and respect. It should encourage exercising initiative within the scope of the commander's intent. However, loyalty and respect must not detract from stating hard truths in departmental assessments.

Before a decision, department members should give honest, independent thoughts and recommendations to their commander, so that the commander can confirm or restructure the commander's visualization. The departments should provide regular updates and recommendations to their commanders, so that they are kept appraised of any situation.

Department members should base recommendations on solid analysis and present them to the commander, even if they conflict with the commander's decision. Independent thought and timely actions by the departments are vital to mission command.

4.3 Seconds in Command/Charge (2IC)

In their efforts to safeguard their business from Cyberspace, the most common failings are in relation to human resources redundancy. In the military, they train their teams so that they have the capacity to ensure that each team member knows and understands the roles and responsibilities of one rank above, and one rank below, theirs. In this way, should a team member be unavailable, the team can continue to operate effectively.

Consequently, the C2 structure requires a hierarchical structure, where there is a delegation of responsibilities from the Commander, through their Deputy/Assistant Commanders, through to the team members. This hierarchical structure encourages effective communication flows between the Commander and their Deputies/Assistant Commanders, with the Deputies/Assistant Commanders knowing that they can go to their Commander for additional resources, should the need arise.

Where your organization has Commander or specialist roles, how do you deal with these personnel being unavailable (e.g. Holidays, Illness, Leaving the business, Promotion, etc.)?

4.4 Training

Appropriate training to ensure that the leaders are tactically and technically competent is an essential component of an effective C2 system. Increasingly, you will see advertisements for job roles that have a list of requirements that are 'as long as your arm'. However, any job candidate that happens to fulfil most of these job requirements will likely be extremely rare (as no two job requirements will be the same) and those that do have all the required skillsets will come with high salary requirements. Additionally, they will be in high demand and may not even be a suitable fit for the business. Consequently, the organization may attempt to make an individual become a company fit, just because the have the requisite skillsets that they are looking for. Inevitably, this leads to things not working out and the individual leaving the business several months later.

With the increased reporting of a shortage of security specialists, businesses need to start to recognize the minimum skillsets required and to formulate a suitable development program, which they can then apply to the candidate that best presents the best fit for the organization.

Through this approach, the organization can engage in a program to mold a suitable candidate into the best shape for the available position. This molding could involve a formal training program of on-the-job, internal, and external training engagements, spread over many months or years. This has three benefits:

1. The individual feels that they are a valued member of staff.
2. The business gets longevity from their staff.
3. An increased feeling of mutual respect is created.

4.5 Information Management (IM)

Advances in business technologies are helping to enhance the organization's IM capabilities and an effective Information Security culture is essential to ensure that the sensitivity of the information is respected and managed in accordance with these levels of sensitivity.

Think of information assets as being the pieces of a jigsaw puzzle. Some of these jigsaw pieces are more influential (e.g. Corner, Edge, Sky, Land pieces) that others and anyone piecing together these pieces need to ensure that they respect how the pieces are inter-linked and that any loose pieces are protected from getting lost, and that the pieces are securely returned to the box when the jigsaw is no longer being played with.

Additionally, once the jigsaw is no longer needed, the box (securely retaining all the pieces of the jigsaw) is securely stored away in a cupboard for secure long-term storage. Finally, once the jigsaw puzzle has been out grown or outplayed *(no legitimate reason/need to retain)* it needs to be appropriately disposed of or destroyed.

There are many similarities between the principles of managing and using a jigsaw puzzle and the management of sensitive information. Careless management can lead to loss of the pieces that are essential for the puzzle.

As much as effective Information Management is important to successful business operations, it is equally important for the C2 system—enabling personnel to efficiently communicate issues of concern between different areas of the business.

It is easy to appreciate that just as the business operations, an effective C2 system is heavily reliant on the completeness, conciseness, accuracy and currency of the information currency and, therefore, effective IM becomes a critical part of both the business and its C2 system.

4.6 Information Systems

The information systems are those business equipment and facilities that support the collection, processing, storage, display and transmission of sensitive business information. To help ensure that these systems remain operational and secure, it is essential that they are supported by an effective Information Security program, where all personnel understand their policies and procedures for the correct use and management. When properly implemented this will aid Commanders to achieve and maintain higher levels of IM value and proficiency.

Should a poorly maintained, managed, or used information system be compromised it could significantly impact your business operations. Think of the information systems being like the mechanical components of a motor vehicle and the fuel being information. Should the petrol tank be punctured or rotted out, or the fuel line be corroded, or blocked, the engine will be starved of fuel and it will come to a grinding halt.

Business Purpose and Capabilities

Your defensive efforts to protect your business' cyberspace environment needs to be a team effort, ensuring that these efforts are aligned with the business context and to ensure that everyone has a sufficient abilities, suitable tools and resources to deliver their roles and responsibilities, in support for the defense of the valued business assets.

Roles

As previously mentioned, the C2 system needs a hierarchical structure with the Commanders in Chief (Board Members) delegating security responsibilities and risk ownership down through the Commanders, 2ICs and team leads.

Where the identified risks exceed the delegated risk appetite levels, there needs to be a process where risk assessment and decision-making process is escalated up the command chain.

Whether an employee works in the backend business operations (e.g. IT Operations, Information Security Management, Risk Management, etc.) or within the frontend business operations (e.g. Sales, Marketing, Administration, Human Resources, etc.), everyone should understand that they have a security responsibility to help protect the organization's cyberspace environment. Consequently, it is essential that businesses embrace and implement a collaborative approach to securing the business operations and data assets:

- Legal, safe, and secure use of the systems and data assets.
- Securely configured and maintained systems architectures.

Systems Architecture

Within many modern digital businesses, they are increasingly reliant on technologies in support of essential business functions and for the processing, storage and transmission of large volumes of data.

The supporting systems architecture should be designed using the principle of least privilege, so that access and connections to more sensitive area of the connecting systems architecture is strictly limited, based upon legitimate needs.

Think of your systems architecture in the same way as the physical architecture designs for a high street bank. By the very design of the layout of a bank, someone walking in off the high street is prevented from walking straight into the bank's vault.

Your organization's systems and network architecture should follow a similar approach, so that the public-facing systems architecture are segregated off from other more sensitive internal IT systems. Any network traffic should be subject to layered defenses to filter access, based upon strict 'need to know'/'need to access' principles.

One example of this would be the placement of perimeter firewalls and internal firewalls, which employ differing and progressively stricter policy rulesets to filter the network traffic.

Additionally, all the supporting systems architecture should be secure configured to ensure that no unnecessary services, ports and protocols (SPPs) are enabled, which would allow an opportunist attacker to circumvent any security protocols at the higher operating system layers, as depicted by the Open Systems Interconnection model (OSI model) (Fig. 2 [14]).

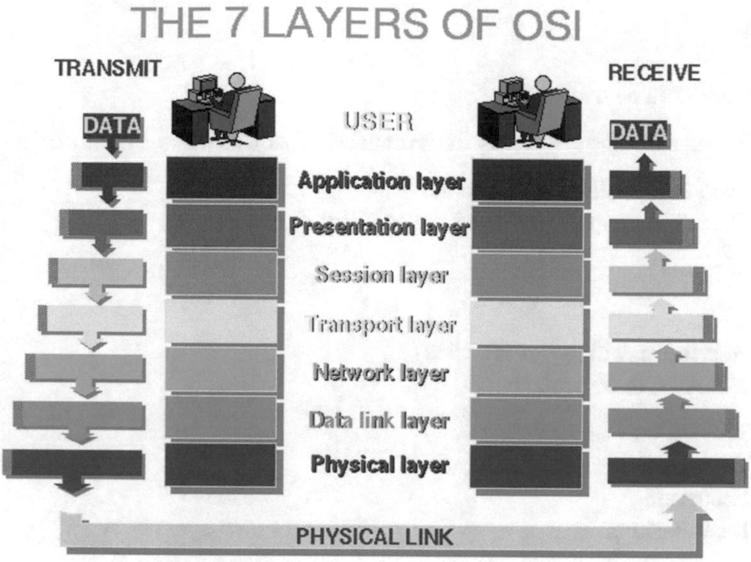

Fig. 2 7-layer OSI model

4.7 Information Systems and Information Management Activities

With the increasing business dependence on data assets, the evolution of many country data protection/privacy laws and the growth in the cyber-criminal's exploitation and monetarization of data assets, it has never been more important to have robust and secure information systems and information management activities.

Notwithstanding the security perspective, data quality is an essential aspect of good business:

- **Quality In, Quality Out**.

Collect

- How is data collected?
- Is this data collected for legitimate legal or business reasons?
- Do you have the data subject's consent to collect this data?

Process

- How is this data processed?
- Where is this data processed?
- Why is this data being processed?

Display

- Can this data be overlooked?

  - Think about the sensitivity of the data being viewed and the local environment.

- Are you in a public environment?
- Do the people around you need to see this data?

Store

- What is the sensitivity of this data?
- Does it need to be protected?

  - Encryption.
  - Tokenization.
  - Redaction.
  - Data masking.

- Do you know the locations where it is being stored?
- Is access strictly limited to the data stores?
- How long is the data to be stored for?
- Is any data being needlessly stored *(the just in case syndrome?)*.
- What is your data retention schedules?

Transmit

- Do you understand the sensitivity of the data to be shared?
- Do you limit the amount of shared data, to the absolute minimum that is required for the specific business task?
- Does the intended recipient have approval to receive this data?
- When considering transmitting sensitive data, do you have the appropriate encryption capabilities?
- Before sending any sensitive data, have you confirmed the recipients
- Have you satisfied yourself that the recipients will afford the same levels of protection to this data, as you would expect and that they will secure dispose of/delete the data when they no longer required?
- Have you considered whether you might need to track the delivery and receipt of the sensitive data?

4.8   Policies and Procedures

Your policies and procedures are essential components of a C2 system, to clearly articulate the company's rules and operating procedures that are required for the continued security of the business' information systems and data assets.

Formal Policies

Much like driving on a public highway, each employee needs to have read and understood the 'Rules of the Road', which should be made available to them. Onboarding a new member of staff should be made to read and understand any policies that are applicable to their job role and existing employees should be required to periodically refresh their understanding of their applicable policies.

The policies should be kept current and be reviewed annually and be updated to reflect any adjustments to the threat landscape or any changes to the business operating model.

Formal Procedures

Once again, using the driving analogy, all employees should be made aware of the step-by-step actions that are needed for the continue safe and secure business operations.

- Think along the lines of:

  - Mirror.
  - Signal.
  - Brake.
  - Gear.
  - Mirror.
  - Maneuver.

Formal procedures help to standardize employee actions and to help significantly reduce the risks to the business.

4.9 Departmental Standing Operating Procedures (SOPs)

The department managers, acting in the role of Commander, should develop local department SOPs that are written in a clear, concise, and suitable language which makes it easier for the Commander's team members to understand.

These SOPs should be owned, managed and enforced by the Commander, and should support the organizational policies and procedures.

4.10 Asset Management

An asset is defined [15] as:

> A *major application, general support system, high impact program, physical plant, mission critical system, personnel, equipment, or a logically related group of systems.*

The first part of asset management requires an organization to identify and understand which businesses are important to the business and to ensure that they are effectively management across their lifecycles. By the very nature of cyberspace operations, businesses need to respect and respond to these dynamic environments, ensuring that the risks to the important assets are identified and any remediation activities are prioritized, based upon the potential impacts to the business.

Equipment

Frequently misunderstood or underappreciated, knowing which equipment can impact your business operations can be wide ranging and it is important to look at each important business process and to understand which equipment is essential to the continual operations.

Facilities

During the COVID19 pandemic, many businesses experienced the disruption and impact that occurred when their business facilities were made unavailable to use, with many organizations having to quickly formulate more flexible remote working operations.

However, not having the relatively reassuring sanctity of a dedicated workplace can considerably increase the inherent risk and, as a result, when creating risk scenarios for your business you should include all the perspectives related to those assets (IT systems, People, Equipment, Facilities, Buildings, etc.) which provide essential support to the important business operations.

5 Recommendations

As you will have read, the effective defense of an organization's cyberspace requires a team approach, involving widespread representations from valued parts of the business in support of a military style C2 structure.

Ultimately, when things go wrong or the business is the victim of a cyber-attack, it will the Commander In Chief who will be held accountable for the actions (or lack of actions) that occurred within the business. Consequently, having an effective and supportive hierarchical structure to help maintain secure and operational processes has never been more important.

Rather than having a single or isolated approach, it is highly recommended that companies create a formal C2 structure where departmental representatives have delegated security responsibilities, which create an effective security culture that permeates throughout organization.

The components of an effective C2 requires the harmonization between 'Frontend' and 'Backend' operations, involving the supporting IT Systems, Supporting equipment & facilities, People and Processes.

All these components should be configured and operated with secure by design/secure by design so that that the considerations of managing the dynamic nature of cyberspace operations are kept at the forefront of business operations.

Integral to the development of a successful C2 system is the establishment of effective communication channels which flows both horizontally ('Top-Down' and 'Bottom-Up'), as well as laterally (inter-departmental), to ensure that good practices are communicated throughout the organization *(creating an intelligence network)*.

Additionally, it is important that the 'Rules of the Road' are documented in company policies and procedures, and departmental procedures, with all personnel having read and understood their applicable 'Rules of the Road'.

Complementary to the C2 system is the establishment of a formalized training program, delivered at both the company and departmental level. This will help to ensure that personnel, no matter their role, understand the extent of their security responsibilities and obligations that are applicable to their job role.

6 Conclusion

All too often businesses are relying on a small number of security specialists to try and manage the risks and security needs for ever-growing digital and data environments.

The result being that too many plates need to be kept spinning by far too few personnel. Investment in the development of a C2 style structure in support of the defense of an organization's cyberspace operations can provide a significant return on investment, with more security aware 'eyes and ears' sharing the obligations to help defend a business from opportunist attackers and to maintain resilient systems.

The military have successfully employed the C2 model in defense of their mission critical operations, both at home and during overseas deployments and such approaches can provide significant benefits to the modern digital business.

Unfortunately, the number of digital businesses that have recognized the benefits of the C2 style approach to defending their valued operations and data assets are extremely limited, which increases the risks of bad practices that might result in a breach of confidentiality, integrity or the unavailability of essential services, systems or data. Such and event (or several events) could result in a significant impact for the business, be that reputational or financial.

References

1. www.etymonline.com (n.d.) cyberspace | Search Online Etymology Dictionary. [online] https://www.etymonline.com/search?q=cyberspace&ref=searchbar_searchhint. Accessed 18 Feb 2021
2. Merriam-webster.com (2020) Definition of INTERNET OF THINGS. [online] https://www.merriam-webster.com/dictionary/Internet%20of%20Things.

3. www.etymonline.com (n.d.) digital | Search Online Etymology Dictionary. [online] https://www.etymonline.com/search?q=digital. Accessed 18 Feb 2021
4. www.mckinsey.com (n.d.) What "digital" really means | McKinsey. [online] https://www.mckinsey.com/industries/technology-media-and-telecommunications/our-insights/what-digital-really-means
5. www.etymonline.com (n.d.) leadership | Search Online Etymology Dictionary. [online] https://www.etymonline.com/search?q=leadership&ref=searchbar_searchhint. Accessed 18 Feb 2021
6. studylib.net (n.d.) leadership—Royal Air Force. [online] https://studylib.net/doc/8180724/leadership---royal-air-force. Accessed 21 Feb 2021
7. Simplilearn.com (2015) What's the difference between leadership and management? [online] https://www.simplilearn.com/leadership-vs-management-difference-article#:~:text=Managing%20People%3A
8. www.etymonline.com (n.d.) strategy | Search Online Etymology Dictionary. [online] https://www.etymonline.com/search?q=strategy&ref=searchbar_searchhint. Accessed 18 Feb 2021
9. www.etymonline.com (n.d.) governance | Search Online Etymology Dictionary. [online] https://www.etymonline.com/search?q=governance&ref=searchbar_searchhint. Accessed 18 Feb 2021
10. Rand.org (2019) Military Command and Control. [online] https://www.rand.org/topics/military-command-and-control.html. Accessed 11 Oct 2019
11. MSW (2015) Roman Command and control in action. [online] Weapons and Warfare. https://weaponsandwarfare.com/2015/08/03/roman-command-and-control-in-action/. Accessed 24 Feb 2021
12. www.encyclopedia.com (n.d.) Command and Control Systems | Encyclopedia.com. [online] https://www.encyclopedia.com/social-sciences-and-law/sociology-and-social-reform/sociology-general-terms-and-concepts/command-and-control-systems. Accessed 22 Feb 2021
13. www.globalsecurity.org (n.d.) FM 6-0 Chapter 1, Command and Control. [online] https://www.globalsecurity.org/military/library/policy/army/fm/60/chap1.htm.
14. Network Engineering Stack Exchange (n.d.) ip—Question about the 7 layers of OSI Model. [online] https://networkengineering.stackexchange.com/questions/42058/question-about-the-7-layers-of-osi-model?noredirect=1&lq=1. Accessed 10 Mar 2021
15. Committee on National Security Systems Committee on National Security Systems (CNSS) Glossary (2015) [online] https://www.serdpestcp.org/content/download/47576/453617/file/CNSSI%204009%20Glossary%202015.pdf. Accessed 10 Mar 2021

Towards an IoT Community-Cluster Model for Burglar Intrusion Detection and Real-Time Reporting in Smart Homes

Ryan Singh, Haider Al-Khateeb, Gabriela Ahmadi-Assalemi, and Gregory Epiphaniou

Abstract The systematic integration of the Internet of Things (IoT) into the supply chain creates opportunities for automation in smart homes from concept to practice. Our research shows that residential burglary remains a problem. Despite the paradigm shift in ubiquitous computing, the maturity of the physical security controls integrating IoT in residential physical security measures such as burglar alarm systems within smart homes is weak. Sensors utilised by burglar alarm systems aided by IoT enable real-time reporting capabilities and facilitate process automation which can be innovatively employed to increase security resilience and improve response to a burglary in smart homes. We research key-related methods of proposed security of home alarm systems and introduce an IoT Burglar Intrusion Detection (I-BID) solution, a new privacy-preserving alarms system with multi-recipient real-time reporting of intrusion in smart homes. Our approach is demonstrated on a developed and tested prototype artefact. The experimental results reveal that the proposed technique reliably detects intrusion, achieves real-time reporting of a home intrusion to multiple recipients autonomously and simultaneously with a high degree of accuracy. The key strength of our technique is its scalability to a community-cluster model as a burglary security mechanism.

Keywords Internet of Things · Monitor · Detect · Alert · Police · Incident response · Intrusion · Sensor · Smart home · Unauthorized access · Burglary

R. Singh · H. Al-Khateeb (✉) · G. Ahmadi-Assalemi
School of Mathematics and Computer Science, Wolverhampton Cyber Research Institute (WCRI), University of Wolverhampton, Wolverhampton, UK
e-mail: H.Al-Khateeb@wlv.ac.uk

G. Epiphaniou
University of Warwick, WMG Group, Coventry, West Midlands, UK

53

1 Introduction

The Internet of Things (IoT) is a novel paradigm integrating physical devices and sensors with digital capabilities of network technologies [1] creating new opportunities for performance and functionality enhancement. According to [2], IoT fused with the 5th generation mobile network's (5G) ubiquitous infrastructure is expected to have a "massive impact on society and business bringing about societal and economic opportunities for everyday connected objects and innovative applications across several smart sectors including smart homes". Although, according to [3] smart homes are in the earlier stages of development compared to other smart city sectors, safety-related applications in smart homes, are one of the forward-looking IoT driven innovations to help address problems in the physical security control realm.

Office of National Statistics (ONS) reporting shows that despite an overall decrease of 4% to 417,416 in burglary offences recorded by police across England and Wales (excluding Greater Manchester figures) in June 2019, of which 291,816 were categorised as residential burglary [4]. Although burglaries reported by the police showed a slight decrease as reported for the period ending in June 2019, these offences had shown a rise in recent year by 6% in the year ending March 2018 and 3% in March 2017 [4]. In the year ending September 2020, ONS reported a 20% drop in burglary compared with the previous reporting period, however, it is noteworthy that this drop is largely attributed to the lockdown restrictions and fluctuation in police recorded crime during this period [5], and residential burglary remains a problem. It is worth noting, that burglary is a crime considered to be well reported by the victims and well recorded by the police.

According to the ONS Crime Survey for England and Wales (CSEW) dataset [6] covering the period between March 2010 and March 2020, the dominant method of intrusion by criminals has consistently been the front door, followed by the back door, with crime prevalence during the week compared to the weekend and the evening or night period over mornings or afternoons with the dominant method of entry being forced locks. Apart from a monetary impact and damage to property, this type of crime considerably affects people in other ways including physically through the use of violence and emotionally [6], which is putting residents including vulnerable groups of people at additional risk.

Visible and audible home intruder alarms are considered a deterrent to a potential intruder. The traditional approach to home intruder alarms in the UK typically consists of audible, visible and remotely monitored alarms through an Alarm Receiving Centre (ARC) by a National Security Inspectorate (NSI) or Security Systems and Alarms Inspection Board (SSAIB) regulated provider. Depending on a plan, response varies from a keyholder only to combined with police response in the event of a confirmed alarm. The police response is governed by the National Police Chief's Council (NPCC) [7].

Standalone house alarms are not monitored by an ARC, therefore have no direct access to the police backed response. A final factor to consider is marginalised groups, examples include the rental market households require the agreement of the landlord and centrally maintained security systems can be potentially unaffordable for lower-income households.

The emerging concept of smart burglar alarm systems leverages IoT to address the problem of the traditional human-in-the-loop approach of intrusion detection in homes. In this article, the problem of physical security controls in smart homes is discussed alongside the concept of an IoT-based community-cluster model for physical security control in smart homes and real-time reporting in the prevalence of a detected intrusion. We discuss the key advancements in technology within homes and examine recent physical security protective systems. Furthermore, we present a system design, construct and evaluate a prototype artefact for a novel IoT based home security alarm system.

2 The Advancement of Digital Technology in Smart Homes Security

2.1 The Role of Transformational Technologies

IoT has evolved from the Radio Frequency Identification (RFID) community attributed to Kevin Ashton [1, 8], which focused on tracking the location and status of physical things to the more recently accepted description of converging the physical and digital worlds [1] where the network of many physical objects such as sensors, software, and network connectivity allow objects to exchange and collect specific data over the internet automatically or manually [9]. This continued growth of disruptive technologies leads the way for the next generation of IoT enabled ubiquitous sensing systems in smart homes. Examples include smart appliances, and systems including environment, utilities, entertainment and security control systems which compared to the more traditional methods of control mechanisms can be operated from anywhere utilising cloud technologies [10, 11].

As a rule of thumb, the proliferation of IoT sensors and devices requires modern security controls and data sharing models. Luckily, we have seen an evolution in related transformational technologies such as Blockchain. Blockchain can prove to be a method to ensure a much more joined-up and integrated approach to data sharing including when processing sensitive data such as clinical trials [12, 13]. We think this technology can be utilised to further support this type of collaborative systems.

2.2 Security Alarm System Approaches in Smart Homes

An Arduino-based home intruder alarm system [14] utilised laser light. The method consisted of a laser to detect movement and a buzzer to sound after motion has been detected by the laser. The switch was used to activate and deactivate the laser and the Arduino microcontroller to make the system operate, function and work. The main concept of this idea involved the laser security system is fully activated by the switch which the resident had to activate. The laser would trigger if an object passed between the laser light and light-dependent resistor, causing the buzzer to go off and alarm the resident, indicating an intruder. According to [14] the test results from the prototype system achieving an overall 80% success in detecting intruders by tripping the laser.

The authors in [15] proposed a Rochelle salt integrated Passive Infra-red (PIR) sensor Arduino based intruder detection system. The method consisted of a PIR sensor for suspicious movement detection and a Rochelle salt used as another sensing element to reduce the false alarm rates. Additionally, an internal buzzer was implemented to alert the resident of an intruder after a motion was detected within the home. The main concept of this idea is that when an intruder trespassed into a home, the PIR sensor would detect movement, which caused the alarm to go off and sound, alerting the resident of a burglar. According to the authors [15], the Rochelle salt improved the system's performance, achieved a more efficient detection and decreased the false alarm rate.

Another study [16] proposed a unique IoT Arduino based door unlocking security system with real-time control for a home. The system consisted of an Arduino microcontroller, RFID reader and tag, wireless transmitter and receiver, a database and a webpage. The main concept of this idea involved improving the security of homes, offices, laboratories and libraries, by monitoring authorised individuals' movements in and out of a building thus preventing unauthorized individuals from gaining access. To gain entry, an authorised individual would have to swipe the card on an RFID reader in order for the system to authorise access. Once authorised, the information was transmitted wirelessly and stored in an online database. Access logs registered successful and denied access and the date stamp, which could be monitored by an authorised individual for a potential unauthorised entry. According to the authors [16], the door could be remotely controlled from anywhere in case of emergency. In addition, the authors asserted that their proposed security system is efficient, secure with a capability to remotely monitor the movement status including unauthorised entry attempts in addition to the exit of authorised individuals.

Likewise, [17] proposed a unique IoT home security system. The system consisted of two modules. Module one included a 4 × 4 matrix keypad, Arduino Uno microcontroller, potentiometer, Liquid Crystal Display (LCD), Sonoff SV, electromagnetic lock and a 12 V (Volt) power bank. The second module consisted of Global System for Mobile Communications (GSM), Infrared Camera, PIR motion sensor, buzzer, battery backup. The main concept of this idea utilised the first module to control the door locking and unlocking with the electromagnetic lock. The LCD showed the status of whether the attempted door access was authorised or unauthorised after the password was entered by the individual wanting to gain access via keypad. If the password matched, then a signal was generated and sent to the Arduino microcontroller to allow the door to be unlocked. The second module was utilised and activated when the residents were away from home. Once the second module was

Table 1 Comparison of different approaches to home intrusion security systems

Study	Does the proposed system alert the owner?	Does the proposed system have the capability to alert the police or an ARC?	Is the proposed system designed in a way to protect vulnerable and marginalised groups?
[14]	NO	NO	NO
[15]	NO	NO	NO
[16]	NO	NO	NO
[17]	YES	NO	NO
This study	YES	YES	YES

activated it worked in parallel with the first module. The security was enhanced by the PIR motion detector with the alarm sounding, infrared image capture and an SMS message sent to the owner via a GSM module alerting the owner of an intruder.

The comparison of the different approaches to home security systems outlined in this section is summarised in Table 1. Whilst the investigated Arduino-based home intrusion detection systems dealt with the problem of sounding an alarm to alert the residents, only one of the investigated studies presented a feature to alert the owner utilising a GSM module. The problems that continue to persist in protecting smart homes are the absence of convergence between the physical components, the digital domains and people, lack of real-time interactive risk-based monitoring and protection of vulnerable and marginalised groups. The solution to this problem is presented in the concept of an IoT community-cluster model for burglar intrusion detection and real-time reporting in smart homes.

3 System Design

We present the IoT Burglar Intrusion Detection (I-BID), the I-BID home security alarm system aims to converge the physical, digital and people realms, achieve real-time monitoring and alerting, and design a solution in a way that is inclusive of vulnerable and marginalised groups. How these aims are achieved is described in detail in the remainder of this section, by producing a design schematic, unique pseudocode and the related data flow diagrams. The key strength of our system is the novel design which enables the communication process in real-time to multiple recipients autonomously and simultaneously compared with proposed systems in other studies.

3.1 I-BID Schematic Design and the Dataflow

The main system components which are presented in Fig. 1, consist of the HC-S04 motion sensor, the WiFi shield and the Arduino Leonardo microcontroller board. The WiFi shield is wired to the Arduino Leonardo microcontroller board to connect to the internet and communicate with online services to establish the wireless capabilities for the SMS, and it is wirelessly connected to a home router, which creates the capability to autonomously and simultaneously interact in real-time with multiple recipients. This could include the owner, community-based smart-neighbourhood-watch type buddy, ARC and including law enforcement. That said, although the presented concept creates this capability, the authors acknowledge that in the UK the police trigger is strictly regulated [7]. The data flow diagram highlighting the key system processes is shown in Fig. 2.

The connections were made to their appropriate inputs in approximately four steps to wire the system:

- The first step was to connect the Ground (GND) from the ultrasonic sensor t GND on the Arduino Leonardo microcontroller.
- The second step was to connect the Voltage Common Collector (VCC) to 5 V on the Arduino Leonardo microcontroller.

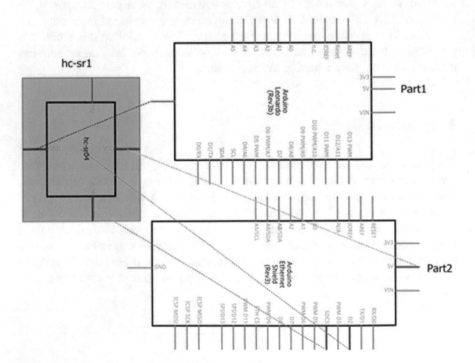

Fig. 1 Schematic diagram of the I-BID home security alarm system

Fig. 2 Dataflow the I-BID
home security alarm system

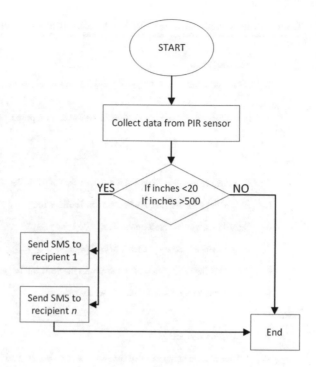

- The third step was to connect the echo pin, which will be directly connected to pin 4.
- In the fourth and final step, the trig pin was connected directly to pin 2 on the Arduino Leonardo microcontroller, which allows ultrasonic waves to be sent out that allow motion to be detected.

3.2 Pseudocode

The pseudocode for the I-BID home security alarm system is outlined in Table 2. The ultrasonic sensor sends the SMS message via WiFi, in this case, the WiFi shield. The sensor communicates with a service called Temboo which utilises a Twilio account where the numbers are set for the SMS recipients, which then sends a chore SMS to the selected recipients' phone numbers.

The following steps are described by the pseudocode:

- Step 1: initially activates the system.
- Step 2: the ultrasonic sensor calculates the distance from the sensor to the closest object in front to get ready for detection.
- Step 3: prints a comment that indicates when the sensor has finished measuring the distance and is ready to detect.
- Steps 4–9 is processed within a loop:

Table 2 The pseudocode for the I-BID home security alarm system

Start

    1.   Activate system

    2.   Ultrasonic Sensor will calculate distance from the sensor to the closest object in front automatically

    3.   After Ultrasonic Sensor has measured its distance, then println "system ready"

LOOP

    4.   IF inches are less than 20 or/II if it is greater than 500 / motion is detected

    5.   Println "Intruder detected, sending the SMS notification"

    6.   Communicate or call the Temboo client

    7.   Set the Temboo account credentials/app key name

    8.   Set choreo inputs/set phone numbers of who the SMS will be getting sent to

    9.   Send SMSChoreo to interlinked phone numbers.

While

    10.   While SMS is sending

    11.   Tell the process to run and wait for results to see whether or not the SMS has been sent?

IF

    12.   If return code is 0,

    13.   Serial. Println "SMS SUCCESSFULLY SENT"

    14.   Success = True

} Else {

    15.   If no return code of 0, then SMS has not been sent.

    16.   Serial, println "SMS has not been sent"

    }

}

    17.   SendSMSchoreo.close

    18.   Delay for 10 seconds

    }// end if statement

}//End loop

- Step 4: If the distance deviates from the sensor measurement completed in step 2, then motion has been detected.
- Step 5: prints "Intruder detected", sending SMS alert after motion is detected.
- Step 6: calls the Temboo service which allows access to the Twilio account so that a choreo SMS message is sent to the selected recipients simultaneously.
- Step 7 and 8: to successfully send the SMS, phone numbers and an app key name have to be added to the algorithm. More specifically, these numbers are the numbers that are set up on the Twilio service.
- Step 9: after Temboo and Twilio have communicated with each other, the SMS message is sent via the carrier network to the selected and added phone numbers after motion is detected by the sensor.

- Steps 10 and 11: is where, while SMS is sending to the pre-set phone numbers, instruct the process to give feedback on results to find out if SMS was successfully sent.
- Steps 12, 13, 14: consist of telling the process, if SMS has been sent, then return 0, with
- Steps 15 and 16: involving else if 0 has not been returned, then prints "SMS has not been sent".
- Steps 17 and 18: by ending the process of sending the SMS, ends the overall process.

4 Constructing and Testing the I-BID Home Security Alarm Artefact

During the construction phase, component-level testing was carried systematically to analyse that components within the system, Fig. 3 are operating and functioning according to the design criteria. This was achieved by using a test-table method while

Fig. 3 The sensor has been connected to the WiFi shield

constructing, Table 3. The table consisted of six columns: "Test ID", "Test Reason", "Expected outcome", "Actual outcome", "Pass/Fail", "What will be investigated to fix the problem, and how was the problem fixed".

The code is shown in Fig. 4 outlines the steps of when the sensor detects movement, the sensor communicates with Temboo service, which communicates with Twilio and tells Twilio to use the specified phone numbers to send an SMS to the selected phone numbers alerting them of intrusion. Twilio details, specifically the number that will be used to send an SMS message to the recipients, has been activated and implemented within the code, Fig. 4, and uploaded to the Arduino Leonardo microcontroller board.

The test shown in Fig. 5 visualises tests 3 and 3.1, Table 3. In this test, we examine if the distance is measured correctly between the sensor and the wall, and the motion is detected in-between. This was achieved by placing an object, hand in this case, in front of the sensor, to test if the sensor triggers the code and the SMS is sent to the recipients' phones alerting the recipients of the detected intrusion in real-time.

The test shown in Fig. 6 visualises test 4, Table 3. In this test, we examine that test of sending and receiving the SMS was positive, and expected results were achieved. After the object, hand, in this case, is placed in front of the ultrasonic sensor, An SMS text message was successfully received with the expected body of the message "Intruder has been Detected!!!!". Both messages were received simultaneously and autonomously at the same time, which also achieves a key aim of this project.

5   Evaluation of the Prototype Artefact

The prototype artefact's performance was validated against the design criteria. Two smartphones shown in Fig. 7 below are placed within the system to visualise the recipients' phones ready to receive an SMS alert after motion is detected by the sensor.

As soon as the home entry point was breached, the door, in this case, the sensor above the door has detected motion and sounded off the alarm, and the sensor that is connected to the WiFi shield facing directly towards the door has also detected that motion, and successfully sent an SMS utilising the WiFi directly to the designated recipients' mobile phones autonomously and simultaneously in real-time, Fig. 8.

We produced enhancements to the home security alarm systems compared to the methods of previous studies, Table 1. The improvement was achieved by converging the physical components with a unique algorithm and leveraging IoT to alert residents in real-time when an intrusion in their home is detected. The new capabilities around alerting multiple recipients autonomously and simultaneously in real-time have been tested and validated empirically iterating through the design phases and the final artefact. More on the collaborative approach of I-BID will be discussed in the next section.

Table 3 Functionality testing of the I-BID home security alarm system

Test ID	Test reason	Expected outcome	Actual outcome	Pass/Fail	What will be investigated to fix the problem, and how was the problem fixed
Test table 1					
ID 1	To test if the WiFi shield is successfully connected to the personal router	Connected	Connected	PASS	N/A
ID 2	To test signal strength for connection strength and reliability between WiFi shield and personal router	Signal strength between 50 and 60%	62–68%	PASS	N/A
Test table 2					
ID 3 (First test run for sensor)	Test the sensor to see if the distance is measured/detects correctly between the sensor and the wall, and motion in-between	Distance to be measured continuously and only stop as soon as hand/motion is placed in front of the sensor with an output saying "intruder has been detected!!!!" in the serial monitor	A repeated message in serial monitor saying "intruder has been detected!!!!" without anything being placed in front of the sensor	FAIL	**Investigation** – Bugs in code – Wrong wiring connections – The incorrect distance set within the code – Faulty hardware/sensor
ID 3.1 (second test run for sensor)	Test the sensor to see if the distance is measured/detects correctly between the sensor and the wall, and motion in-between	Distance to be measured continuously and only stop as soon as hand/motion is placed in front of the sensor with an output saying "intruder has been detected!!!!" in the serial monitor	Distance measuring correctly and says "intruder has been detected!!!!" within serial monitor when hand/motion is placed in front of the sensor	PASS	**Problem found**: – Set distance was incorrect within sketch/code **How the problem was fixed**: – Distance changed to correct distance

(continued)

Table 3 (continued)

Test ID	Test reason	Expected outcome	Actual outcome	Pass/Fail	What will be investigated to fix the problem, and how was the problem fixed

Test table 3

Test ID	Test reason	Expected outcome	Actual outcome	Pass/Fail	What will be investigated...
ID 4	Testing if SMS is sent to both recipients after the ultrasonic sensor detects motion	SMS sent to: *Recipient 1* "Intruder has been Detected!!!!" *Recipient n* "Intruder has been Detected!!!!"	SMS received by: *Recipient 1*—"Intruder has been Detected!!!!" *Recipient n*—"Intruder has been Detected!!!!!"	PASS	N/A

6 How Can the I-BID Home Security Alarm Improve Homes Security Collaboratively

Traditional home security alarm systems secure homes individually relying on human intervention. However, this is no longer sufficient, and we argue that IoT-based real-time security mechanisms are required as a forward-looking approach. This assertion is supported by the following survey [18] in which Citizen Advice aimed to gain an understanding of consumer protection expectations of what the consumers would expect to be a norm within smart homes. The participants responded that manufacturers of smart locks should identify unusual behaviour proactively and alert the consumer as a burglary preventative measure.

We illustrate parallels with the field of cybersecurity. According to [3, 19] authors assert that despite digitalisation and transition from traditional to an IoT enabled practice, there appeared little evidence of cross-organisational information security sharing and coordination across smart city sectors. Furthermore, practices appeared in silos and lacked cyber-defence collaboration [19]. We argue that a similar phenomenon can be found in smart homes security. Despite the transition from traditional to smart homes leveraging IoT, home security operates in a silo approach. An example being the alerting process which is maintained in a tightly controlled central management structure with a human in the loop. Therefore, advanced and forward-looking solutions are required to support a modern and collaborative home defence approach.

The empirical evaluation of the artefact formed a critical part of demonstrating that the I-BID home security alarm system can be scaled as a community cluster model. Our approach is inspired by the concept of a community-led neighbourhood watch scheme [20], which is potentially a critical factor for vulnerable and marginalised groups.

Fig. 4 The recipients' phone numbers added to the code. Note: Some personally identifiable information has been hidden

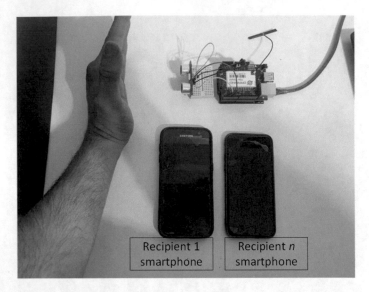

Fig. 5 Placing a hand in front of the sensor to investigate if motion has been detected and SMS has been sent to the recipients

Fig. 6 SMS Intruder alert sent to the recipients' mobile phones successfully in real-time on the left-hand side of the figure and a clearer view of SMS sent on the right-hand side of the figure

7 The I-BID Community-Cluster Model for Burglar Intrusion Detection

Because the I-BID system can, by design, communicate with multiple recipients via appropriate communication channels, it can collaborate with homes within the neighbourhood such as chosen friends and relatives. This could be achieved by extending the artefact to utilise two sensors implemented within the same locations of different homes, as we outline in the I-BID community-cluster model Fig. 9.

Fig. 7 Prototype artefact I-BID home security alarm

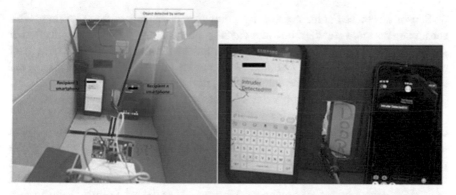

Fig. 8 SMS Intruder alert received by the designated recipients on both smartphones autonomously and simultaneously in real-time on the left-hand side of the figure and a clearer view of SMS received on the right-hand side of the figure

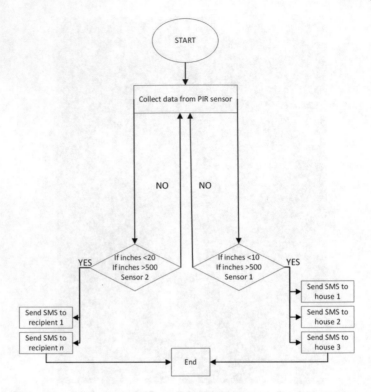

Fig. 9 I-BID community-cluster model for collaborative home security alarm system

Sensor 1 is placed at the home entry point such as outside of the front door. This sensor detects motion at the front door causing the system to collaborate with another system that is part of the community-cluster model within another home. That sensor alerts the resident and the neighbours in the cluster that a motion has been detected by the front door sensor of the collaborating house across the street to indicate that the partnered house is at possible risk of intrusion. This is considered low risk.

Sensor 2 is placed inside the home which will be facing towards the home entry point such as the front door, ready to detect motion and send an SMS alert autonomously and simultaneously to multiple recipients in real-time to their smart-phones alerting them of an intrusion. However, if an SMS is then also triggered from inside that home, it indicates that an intrusion is likely to have occurred. Therefore, the SMS is sent directly to the owner and the model has the capability to send an alert to an ARC or law enforcement alerting all specified recipients of an actual intrusion. That said, we acknowledge that whilst we have designed and demonstrated capability further research, development and empirical testing of the artefact is required.

The homes within the I-BID community-cluster model are federated at the Temboo service layer. To maintain privacy, each home is set up with its own I-BID system consisting of internal and external sensors, a minimum of a pair of sensors, Sensor 1 and Sensor 2. Sensor 1 is placed inside the home entry point and Sensor 2 is placed outside the home entry point. However, the model could support multiple pairs of sensors to be placed at other home entry points e.g. back doors, side doors or windows. The WiFi shield utilises the local wireless router to enable secure communication with the Temboo service which is the layer where the homes are federated into clusters.

For example, in a small cluster of three houses, if Sensor 1 detects motion within home one, then the system will send an SMS message to alert the other two homes in the cluster of a potential intrusion. The same applies to the other two homes, hence, if Sensor 1 detects motion within home three, then home two and one will be alerted by an SMS text message. It is worth mentioning again, that if Sensor 2 detects motion that indicates high risk, then the system will directly send the SMS message alert to the owner and can communicate with multiple specified recipients. Therefore, this could include ARC and law enforcement contacts.

7.1 Cybersecurity Implications of Leveraging IoT in Physical Protection of Homes

IoT based physical security of homes is an evolving field and whilst technologies and innovations that support the delivery of the objectives for homes' physical security are transformational, they also have numerous challenges. Apart from the multifaceted nature of issues such as privacy, ethical challenges and regulatory aspects, one of the key concerns of the IoT based physical protection of homes is cybersecurity.

Consumers desire to use cutting edge technology, comes with various challenges including the proliferation of smart technologies and the lack of understanding of associated risks and vulnerabilities [18]. Due to the threats such as data breaches or data interception, technical security mechanisms including authentication, access control, network and cryptographic protections are needed to protect information when shared systems intercommunicate wirelessly and when cloud services are utilised.

Furthermore, privacy and security need to be maintained between each collaborative home when information is shared to protect the confidentiality and integrity of the data. End-to-end encryption has a role to play as part of the layered defence in-depth approach, for example [21] discusses how the use of end-to-end encryption allows only the communicating users to open and read specific messages and research by [22] suggests that end-to-end encryption decreases the risk of attacks including a man-in-the-middle compromise.

Impact	Probability of the likelihood and impact	Likelihood				
		Very Unlikely	Unlikely	Possible	Likely	Very Likely
	No Impact					
	Low Impact		Sensor 1- outside the home entry point			
	Medium Impact					
	High Impact				Sensor 2 - inside the home entry point	
	Catastrophic Impact					

Legend – Risk Categories

LOW IMPACT	MEDIUM IMPACT	HIGH IMPACT

Fig. 10 A risk assessment matrix for the I-BID community-cluster model

Sensors	What threat could occur	Who will the system alert	Risk rating priority /taking Action
Sensor 1 outside the home entry point	Attempted break-in	collaborated houses in the cluster	LOW
Sensor 2 outside the home entry point	Harm to residents/Aggravated burglary	Capability to alert multiple recipients including ARC	HIGH

Fig. 11 A risk impact matrix for the I-BID community-cluster model

7.2 Risk Assessment

A risk assessment matrix for the sensor thresholds is produced for the collaborative approach, Fig. 10. Sensor 1, being the sensor on the outside of the home entry point, if triggered, is considered unlikely that an intrusion has occurred, Fig. 11. Sensor 2, being the sensor on the inside of the home entry point, if triggered, is considered likely that an intrusion has occurred and the impact of such incident is considered high, which potentially according to Fig. 10 could result in harm to residents.

8 Ethical Considerations of Burglar Intrusion Detection

At an individual level, the desire to use a security mechanism to protect the physical security of homes from potential intruders may give rise to questions of social injustice and equality issues such as affordability and accessibility of technologies and access to effective law enforcement response thus counterbalancing the desired effect of physical home protection as a deterrent of burglary related crimes. According to [23], the police-recorded crime data for England and Wales supports the notion of inequalities and crime including burglary. Furthermore, the author in the same study outlines a link between wealth and security mechanism adoption. Therefore, a commitment to developing a consistent and affordable approach to effective physical home security is required.

At a societal level, according to research by the Institute for the Study of Civil Society [24], a comparison of households with income of above £50,000 and below £10,000 showed that the low-income households were twice as likely to be burgled. Furthermore, this study showed that the households below the £10,000 income threshold were two and a half to three times as likely to live in fear of crime including burglary with further evidence of three and a half times the rate of criminals living in the 20% most deprived areas compared with the same rate in the least deprived areas. The link between wealth and security measure adoption and the level of insurance premiums of the low-income households has been outlined by [23]. Questions could arise if wealth is the ultimate driver of home protection thus mechanisms would be required to make physical home protection available to disadvantaged groups.

9 Conclusion

Can physical security controls of homes be improved in today's era? This study aimed to improve physical security controls within homes regards burglar alarm systems, which could help burglary and intrusion decrease in this day and age. Office for National Statistics [4] stated that there has been an increase of robberies by 12% in 2019 compared to the previous year, and still increasing to this day. Part of this problem was because of weak physical security controls implemented within homes. However, physical security controls that are implemented within homes are not always fit-for-purpose; they fail to alert residents in time of unauthorised access, slow response by police, and false positives. These issues signals why robberies are not decreasing, which is a big current issue that needed to be dealt with and solved. Therefore, our objective was to propose a new and improved IoT based home security alarm system that could offer better protection for homeowners including vulnerable groups of people (e.g. disabled people) from burglary. Nonetheless, reasons for unauthorised access are not always linked to burglary, as it could also include hate crimes and domestic abuse [25], hence the need for a collaborative approach with community support facilitated by technology. The empirical evaluation of the proposed design in this study has formed a critical part of demonstrating that the I-BID home security alarm system can be scaled as a community cluster model. Our approach is inspired by the concept of community-led neighbourhood watch schemes, which is potentially a critical factor for vulnerable and marginalised groups.

References

1. Greer C, Burns M, Wollman D, Griffor E (2019) Cyber-physical systems and internet of things. https://doi.org/10.6028/NIST.SP.1900-202
2. Ahmadi-Assalemi G, Al-Khateeb HM, Maple C, Epiphaniou G, Alhaboby ZA, Alkaabi S, Alhaboby D. Digital twins for precision healthcare. In: Cyber defence in the age of AI, Smart societies and augmented humanity, p 133. https://doi.org/10.1007/978-3-030-35746-7_8
3. Ahmadi-Assalemi G, Al-Khateeb H, Epiphaniou G, Maple C (2020) Cyber resilience and incident response in smart cities: a systematic literature review. Smart Cities 3(3):894–927. https://doi.org/10.3390/smartcities3030046
4. Office for National Statistics (2020) Crime in England and Wales: year ending June 2019, p 66
5. Office for National Statistics (2021) Crime in England and Wales: year ending September 2020
6. Office for National Statistics (2020) Nature of crime: burglary dataset, ONS, ed
7. National Police Cheifs' Council. National Police Cheifs' Council Security Systems Policy. https://www.policesecuritysystems.com/national-police-chiefs-council-security-systems-policy. 09 May 2021
8. Ashton K (2009) That 'internet of things' thing. RFID J 22(7):97–114
9. Patel KK, Patel SM (2016) Internet of things-IOT: definition, characteristics, architecture, enabling technologies, application & future challenges. https://doi.org/10.17148/IARJSET.201 8.517
10. Asadullah M, Raza A. An overview of home automation systems, pp 27–31. https://doi.org/ 10.1109/ICRAI.2016.7791223
11. Modarresi A, Sterbenz JPG. Towards a model and graph representation for smart homes in the IoT, pp 1–5. https://doi.org/10.1109/ISC2.2018.8656928
12. Jahankhani H, Kendzierskyj S, Jamal A, Epiphaniou G, Al-Khateeb H (2019) Blockchain and clinical trial: securing patient data. Springer
13. Ersotelos N, Bottarelli M, Al-Khateeb H, Epiphaniou G, Alhaboby Z, Pillai P, Aggoun A (2021) Blockchain and IoMT against physical abuse: bullying in schools as a case study. J Sens Actuator Netw 10(1):1
14. Arjona AB, Bautista PKM, Edma JE, Martel MIJP, Octavio EDN, Balba NP. Design and implementation of an Arduino-based security system using laser light
15. Saini SS, Bhatia H, Singh V, Sidhu E. Rochelle salt integrated PIR sensor arduino based intruder detection system (ABIDS), pp 1–5. https://doi.org/10.1109/ICCCCM.2016.7918228
16. Nath S, Banerjee P, Biswas RN, Mitra SK, Naskar MK. Arduino based door unlocking system with real time control, pp 358–362. https://doi.org/10.1109/IC3I.2016.7917989
17. Kumar R, Mittal P (2019) A novel design and implementation of smart home security system: future perspective. Int J Appl Eng Res 14(2):363–368
18. Traverse (2018) The future of the smart home: current consumer attitudes towards smart home technology. https://www.citizensadvice.org.uk/Global/CitizensAdvice/Energy/Smart% 20homes%20final%20report%20(new%20Traverse%20logo).pdf. Accessed 13 May 2021
19. Ahmadi-Assalemi G, Al-Khateeb HM, Epiphaniou G, Cosson J, Jahankhani H, Pillai P. Federated blockchain-based tracking and liability attribution framework for employees and cyber-physical objects in a smart workplace, pp 1–9. https://doi.org/10.1109/ICGS3.2019.868 8297
20. Network NW (2020) Neighbourhood watch national crime community survey 2020. https:// www.ourwatch.org.uk/sites/default/files/documents/2021-01/Neighbourhood%20Watch% 20National%20Crime%20and%20Community%20Report%202020.pdf. Accessed 13 May 2021
21. Nabeel M. The many faces of end-to-end encryption and their security analysis, pp 252–259. https://doi.org/10.1109/IEEE.EDGE.2017.47
22. Thomas M, Panchami V. An encryption protocol for end-to-end secure transmission of SMS, pp 1–6. https://doi.org/10.1109/ICCPCT.2015.7159471
23. Newburn T (2016) Social disadvantage: crime and punishment. In: Social advantage and disadvantage, pp 322–340

24. Cuthbertson P (2018) Poverty and crime: why a new war on criminals would help the most poor
25. Alhaboby ZA, Al-Khateeb HM, Barnes J, Jahankhani H, Pitchford M, Conradie L, Short E. Cyber-disability hate cases in the UK: the documentation by the police and potential barriers to reporting. Cybersecurity, privacy and freedom protection in the connected world, pp 123–133

Socio-technical Security: User Behaviour, Profiling and Modelling and Privacy by Design

Paul Wood

Abstract Security risk to information exists in many forms and guises; being human and non-human, and those directed from inside and outside of an organisation. Traditional mitigation methods have predominately focused upon applying layers of reactive security solutions, triggered by breaches to defined rules and processes. In order to counter the risk to information posed by individuals from within an organisation, behavioural research is increasingly becoming a crucial element in the design of automated security tools to identify and respond to anomalous and malicious behaviour. Accurate user profiles and threat models may enable the effective and efficient deployment of user behaviour analytics systems, which in turn enable the proactive mitigation of threats before they cause harm to people, property, including intellectual property, or premises. The consideration of personal characteristics and role-related behaviours during the design of security systems can protect the privacy rights of individuals, improve business effectiveness and cost efficiencies for organisations. In an increasingly mobile world, next generation and anticipatory security risk systems are needed to respond to the shift that people are becoming the new security perimeter.

Keywords Insider threat · Security risk · Profiling · Behaviour analytics · Privacy

1 People Are the Threat

Personnel security threats are high on geopolitical, cyber-risk management and business organisation agendas. Insiders, in the form of infiltrators, exploited individuals, malicious or inadvertent actors exploit, or have the intention to exploit their legitimate access to an organisation's assets for unauthorised purposes. The pervasive threat posed by people within an organisation has intensified and diversified with the development and increased availability of artificial intelligence (AI) and machine learning (ML). The use of AI by malicious actors could enable them to effectively

P. Wood (✉)
University of Portsmouth, Portsmouth, UK
e-mail: information@emergingrisksglobal.com

© The Author(s), under exclusive license to Springer Nature Switzerland AG 2021
R. Montasari et al. (eds.), *Challenges in the IoT and Smart Environments*,
Advanced Sciences and Technologies for Security Applications,
https://doi.org/10.1007/978-3-030-87166-6_4

hack or socially engineer victims, or to interrogate large data sets in order exfiltrate valuable data [5]. Increased access to AI has enabled threat actors to pursue attack vectors that would have previously been viewed as too high-risk, including sending personalised spear fishing attacks to high-value targets. AI has improved the ability of threat actors to identify targets and to conduct research which improves the effectiveness of attacks. This threat may be further intensified by the anonymity provided by AI which may increase the psychological distance between malicious actors and their victims, resulting in a reduction in empathy and a subsequent increased willingness to carry out a malicious action [5]. Despite this emerging security challenge, insider threats remain to be one of the most challenging security issues that organisations and security executives face, with some believing it to be an area that is not well addressed by existing security systems. Closely related to the cost: benefit assessments that drive the procurement selection of security systems, the growing interest in the concept of 'trust' and its influence upon performance has resulted in the additional search by security leadership teams for the appropriate balance between security and freedom of access. In line with the pursuit of 'Security by Design', proactive and passive security systems and processes have been increasingly considered and adopted by organisations. In this chapter, we will consider insider threats and the socio-technical security solution systems that have been designed to mitigate them.

2 The History of Proactive Security

The recruitment of competent employees is critical to the achievement of organisational objectives and potentially provides a competitive advantage over rivals. Similar to government agencies with access to classified material, commercial organisations may require personnel to undergo screening during recruitment for roles which require access to sensitive and privileged assets. Conducting inquiries into a prospective applicant's background, including reference and criminal records checks may be viewed as standard practice, aiming to mitigate the potential liability associated with any incidents resulting from a failure to investigate the background of an applicant. Although a reference check is a common selection practice, carried out to gather information which can be used to assess a potential candidate's behaviour in employment, early research found that references offered low operational validity in relation to work performance [14], with many employers doubting the veracity of the information gathered from people identified by candidates, who may provide individuals as references more likely to provide positive accounts [12]. A check that could offer a more objective validity than a reference gathered from a former colleague or friend, is a criminal history records check. The adoption of this practice has increased both in the United States of America and the United Kingdom due to improved internet access, which has subsequently made checks more available and cost effective, resulting in criminal background checks being conducted for many employment positions [7, 13]. Despite reference checks and the popularity of criminal background checks, it is commonly accepted that the human factor remains the weakest link in security. In an effort to manage this risk, organisations all around the world have been working hard to create positive and proactive security cultures, within which employees and contractors are informed of threats and empowered to make risk-based decisions, designed to protect information and to mitigate attacks. While educating employees is a crucial element of a security transformation programme, a role-risk based approach to the security monitoring of a workforce is an additional necessity, in order to enable the effective identification and disruption of threats before the cause potentially irreparable damage.

3 Employee Surveillance

Once viewed to be synonymous with the covert activities of intelligence services and law enforcement agencies, open-source intelligence gathering (OSINT) has also become an invaluable resource for commercial organisations. Either conducted directly or through third parties, to collect information related to existing or prospective employees, in addition to the performance of traditional personnel screening and investigative techniques, social Networking sites such as Facebook, LinkedIn and Twitter where users can share and comment on posts, photos, messages, videos, and events, and indicate potential preferences on comments and behaviours, have increasingly been considered by human resources (HR) professionals during recruitment, selection, and pre-employment background checks. Such checks support the assessments of trust and competence, which organisations undertake to assure business performance and the security of assets with which an employee may interact. Through interrogating the information that is openly available, HR professionals and investigators can create personality and behaviour models for individuals, based upon inferences of potential behaviour. Such information can supplement other competency evaluations, in order to assess role suitability and to identify any potential 'red flags'. Warning markers attached to the personal habits and preferences of individuals such as the excessive consumption of alcohol, the use of prohibited drugs, extremist or vulgar behaviour, or involvement in a range of socially questionable activities may concern employers due to the potential reputational risks attached to such choices, as well as the increased vulnerability attached to the individuals involved, which may introduce the risk of exploitation by external threat actors.

Although prospective candidates and existing employees may be asked to permit potential employers' permission to view or potentially access social networking profiles, such practices have increasingly come under scrutiny due to concerns regarding the legality of such monitoring. Although employers might be hoping for the ability to review social networking sites as part of their ongoing security monitoring regime, the unregulated practice of gathering OSINT in order to create individual employee profiles through screening social media sites may raise a number of ethical issues. In addition to concerns regarding privacy, there are limitations to the validity of such information sources due to the potential for the creation of false profiles or the abuse of an individuals' social networking accounts by other actors. Furthermore, employers must also consider the problems associated with 'fake news' and fake accounts [1]. The risk exists that individuals may gain access to the account of another in order to carry out malicious activities, which could negatively impact upon the image and profile of the victim and any associated people or organisations. The ethical and reliability challenges associated with social networking sites should all be considered during both OSINT activities and during the construction of employee profiles.

4 Profiling

Profiling is an analysis process which considers an individual's personality, behaviour, interests and habits, with the intention of making predictions or decisions about them. The UK General Data Protection Regulations Act (The Information Commissioners Office, Article 4 (4), 2018 [29]) defines profiling as "any form of automated processing of personal data consisting of the use of personal data to evaluate certain personal aspects relating to a natural person, in particular to analyse or predict aspects concerning that natural person's performance at work, economic situation, health, personal preferences, interest, reliability, behaviour, location or movements". Information is harvested from a number of sources including internet searches, purchase habits, data gathered from mobile phones, social networks, video surveillance systems and the Internet of Things. People are subsequently categorised into groups according to different behaviours and characteristics in order to determine an individuals' preferences, predict their behaviour and to make decisions about them. The construct of profile records regarding identified traits and tendencies have increasingly been adopted by academic institutes and businesses as a means of predicting employee performance [23] and to identify and predict inappropriate traits, similar to the methods adopted by law-enforcement investigators. Criminal profiling, also referred to as offender profiling, psychological profiling and behavioural profiling, has been described as a process which identifies the personality traits, behavioural tendencies, geographic locations and demographic or biological descriptors associated with an individual [19]. The construction of a criminal profile requires that inferences be made based upon the collection of tangible evidence and signs which demonstrate patterns and trends in the behaviour of an individual. Criminal profiling is conducted by highly trained law enforcement officers and forensic psychologists, who use their expertise to analyse behaviours and the choices made by individuals, that might help to identify motives and patterns. The criminal profiling process involves gathering a plethora of information from crime scene reports, witness and victim statements, and event or autopsy reports. The evidence is assessed in terms of its relevance and accuracy, in order to determine its value to any court proceedings which may result from the criminal investigation. Turvey [32] has described offender profiling as the process of inferring the personality characteristics of individuals responsible for committing criminal acts. It should be noted that criminal profiling does not point to specific offenders, but rather it only provides a broad indication of the type of personality traits which may display a propensity for committing particular offences. Given these limitations, criminal profiling may only be used by law-enforcement in order to reduce the size of suspect lists during criminal investigations which possess little evidence. The modus operandi, or method of operation, displayed by an offender may provide clues about the individual responsible for an act and the personality traits that they may display. This is important as the behavioural choices to perform an action are related to personality types and may provide an indication for the motivation behind a malicious activity. Through examination, an investigator may be able to predict which personality types may hold an increased propensity for carrying out a particular offence. When conducted effectively, this process can enable investigation teams to reduce the size of suspect lists, thereby improving their ability to direct resources appropriately. The methods adopted in profiling follow two key assumptions [17].

(a) The consistency assumption. This holds the view that an offender will display similar behaviour.
(b) The homology assumption. This holds the view that an association exists between offence types and offender backgrounds.

The methods adopted in criminal investigations concerning activities in the physical world can also be used to examine malicious and criminal activity in an electronic cyber environment. While similarities do exist, a number of additional factors should be considered which can pose challenges to investigating cyber related activities. This includes the fact that the users under analysis can be remote based, potentially on different continents [16], in addition to the challenges associated with the methods adopted to ensure an accurate profile is constructed. A consistent and rigorous investigation method should be adopted to avoid the risk of variation in findings and subsequent recommendations [4]. Warikoo [35] suggested a hybrid cyber-criminal profiling methodology which adopts a blend of deductive and inductive investigation methods, including similar inductive profiling methods and statistical analysis, as used by the Federal Bureau of Investigation [16]. This method uses data mining techniques in order to construct pattern detection models and analyses data in order to identify patterns. In an attempt to increase global resiliency to cyber-attacks, such data and cyber-attack records are stored on open-source databases such as the Hackers Profiling Project and the Mitre Att&ck framework [21]. Kirwan and Power [18] have proposed a cyber-criminal profiling process designed to consider technological advances and the development of hacking techniques. The methodology consists of six Profile Identification Metrics, designed to identify and classify an individuals' modus operandi, psychology, and behaviour characteristics.

1. Attack Signature. The initial phase will analyse forensic evidence indicating the nature of attack signature.
2. Attack Method. The cyber-attack method is identified.

3.   Motivation Level. The complexity of an attack indicates the level of motivation
     that an actor possesses. High levels of complexity imply that an attacker is
     motivated and as such may be a persistent risk taker.
4.   Capability Factor. Capability factor considers the tools and capabilities used in
     an attack, which provides an indication of the level of resources available to an
     attacker.
5.   Attack Severity. Attack severity provides an indication of the degree of impact
     an attack has had.
6.   Demographics. In addition to recording a geographic location, this indicates
     the background and related skill sets, funding and motivation attached to threat
     actors.

The profiling process suggested by Kirwan and Power [18] outline a number
of factors that could be considered to profile cyber-criminals through identifying
malevolent actions and enabling further predictions to be made, so as to support
cyber-security operations. In addition to proactive cyber-security processes such
as this, each designed to counter external actors, organisations are increasingly
looking to identify and disrupt insider threats through a number of systems. This may
include monitoring professional activities alongside social interactions, as a means
of identifying inappropriate behaviours and activities, with the aspiration of creating
role related workplace profiles, thus enabling the identification of changes within
employees before they result in harmful consequences to a business or organisation.
Such workplace profiling has been practiced in the past as a means of identifying
and preventing fraud and corruption. Under the UK Fraud Act [30], fraud can be
committed a number of ways;

(a)   Section 2 (fraud by false representation)
(b)   Section 3 (fraud by failing to disclose information)
(c)   Section 4 (fraud by abuse of position).

Fundamentally, workplace fraud is an attempt to deceive another, in order to gain
a benefit. This can include a range of actions, including; theft by employees, expense
fraud, forgery and insider trading. All of the afore mentioned and other potential
breaches, violate an organisations trust and will contain three core elements within
actors;

(a)   An opportunity to commit an action
(b)   A motivation to carry it out
(c)   The existence of a belief within an actor, that they won't get caught.

The debate is ongoing to determine whether personality traits or an emotional
attachment to perceived unfair treatment by an employer, increases the motivation
to commit an act and whether this can be proactively identified. This view may
support activities designed to increase positive association with an employer, in
order to counter any growing retaliatory desire within employees [9, 26]. Although
organisations and wider society may never be able to eliminate malicious activities
completely, different approaches can be adopted in order to influence each element.

Recognising the research finding that employees may be more likely to commit a malicious activity based upon a personal rationalisation that they hold, rather than by an opportunity that exists [36], organisations could benefit through the proactive identification of employee engagement and motivation levels. Pre-screening and in role interviews can identify the presence of beliefs within an individual that potentially indicate whether an employee or candidate is approaching a decision rationalising illicit activities. Counsellors may subsequently identify opportunities to encourage individuals to examine their own thoughts and choices as a means of encouraging self-regulation. This however would not counter individuals, who under the "Bad Apple Theory" may possess a propensity for deviant acts regardless of the perceived environment that they find themselves in [22]. Recognising the financial and resource costs of such independent analysis, high volume enquiries may be preferred. A mass of corporate measurements can be generated through the use of employee surveys, designed to identify trends of role and salary satisfaction, alongside levels of social cohesion and content. Organisations have attempted to focus on influencing employee motivation as a means of reducing the incidence of malicious activities, through striving to improve levels of work: life balance, flexible working, encouraging the growth of social networks amongst colleagues and by offering well-being services, all of which reflects the wider duty of care that organisations hold for employees. This is in addition to the presentation of awards for the achievement of performance goals and through the use of financial incentives, to motivate employees [31]. These efforts are all positive steps towards mitigating the risk that employee dissatisfaction has to increase the potential for employees carrying out an inappropriate or malicious activities [3].

While pre-screening checks are designed to identify evidence of previous activities of concern, they may not be a reliable indicator, particularly for employees that remain within organisations for prolonged periods [20]. This may be pertinent in light of the suggestion of [6] that the majority of criminal actions, in particular theft, is carried out by individuals that have been employed by an organisation for between one and five years. The employees' potential propensity to commit a malicious activity may increase alongside their growing awareness and understanding of how an organisation functions. Although efforts to increase employee motivation may benefit business performance [11], it may not mitigate the potential for malicious activities to take place, so a requirement for a range of internal controls remains in order to harden defences around assets and to deter individuals from conducting such actions, for fear of being detected. Given the ability of some individuals to navigate electronic communication systems with stealth, the need to track activities within these environments exists, if risks are to be effectively identified and mitigated. In addition to monitoring activities, benefits can be experienced through the development of employee user profiles, including improvements to the rate of positive results for the identification of anomalous and potentially malicious behaviour. Although research, including the suggestion made by Appelbaum et al. [2] has suggested that some organisations may view the risk of malicious employee activity as too much of a challenging task to approach, the risk posed by the malevolent or inappropriate activities of an insider can be devastating to the operations and reputation of an organisation, and may sometimes be beyond repair. Such deviant behaviour should not therefore be ignored and will require constant monitoring, in order to counter the ability for employees with privileged access to assets to move sometimes without restraint and with varying degrees of stealth. This can be particularly challenging to monitor when the employee behaviour is related to 'work withdrawal', whereby individuals use paid work time inappropriately, resulting in reduced levels of productivity. This can display itself in a number of ways, such as spending prolonged periods in applications or resources than needed, engaging with social media and open-source sites on the internet, tardiness and absenteeism. The detection and the subsequent response to these and other potentially illegal activities require constant monitoring and data harvesting. AI can then be used to create and apply algorithms to this large quantity of data, which can help to identify correlations between separate datasets. This information can subsequently be used to predict behaviour or to control access to a service. Through the use of artificial neural networks (ANN), artificial intelligence has provided a means of gathering information regarding observed user behaviours and has helped to develop an understanding of why common behaviours present themselves. As ANN are systems which loosely model the brain, they are considered adaptive in being able to adjust behaviours in response to inputs. The logging and analysis of such user behaviours can improve the ability of security analysts to make quick and consistent decisions, particularly when large data sets are involved. The resulting algorithms can be used to respond to user behaviour anomalies in real time after identifying any deviations from baseline activities.

5   User Behaviour Analytics

User Behaviour Analytics or UBA develops an understanding of what a user commonly does during the course of their work or role, through the creation and analysis of logs, recording system activities including the applications launched, network movements and auditing what files are accessed, by who, when and what was done. A comparison is made between logs and identity and access management records, to determine whether the audited access has been authorised and authenticated, and whether this is in accordance with the expectations and requirements attached to a role. The capture of behavioural logs offers an opportunity to observe naturalistic human behaviour uninfluenced by observers, thereby providing an uncensored view of user behaviours and activities. Recognising that the most common and significant threat to information security comes from the behaviour of its internal users, rather than from malware or from malicious external attacks, security systems have been designed to monitor behaviours and to subsequently develop security incident and event management rules to identify suspicious activities. During a period of data collection, UBA will record both authorised and unauthorised activities and the resulting rulesets will enable UBA systems to differentiate between permitted behaviour and unusual or anomalous behaviour which could pose a threat. The installation of a UBA system may be an appropriate step towards reducing the risk resulting from the behaviours of humans and the inappropriate use of information resources and security [33, 34]. The improved employee performance and the appropriate behaviour displayed by system users resulting from the effective implementation of UBA, supports the achievement of the three core functions of protection systems designed to secure information assets; confidentiality, integrity, and availability.

The analysis of user behaviour data offers a means of detecting and investigating complex attacks, comparing log events which represent normal user behaviour with potentially suspicious activities. This tool can improve the ability to detect the malicious activities of sophisticated attackers, who consciously move with stealth across an internal network. As the manual analysis of recorded log events can be both complex and time consuming, automated systems may support the drive towards 'always on' security systems and personnel screening. Automated tools are required that effectively differentiate malicious insider activities from normal business activities before a malicious activity can result in harm. In order for this to be possible, a system must be installed and implemented which provides meaningful descriptions of log events, including session times, duration, location and systems used. Grouped together, comparisons can be made to predefined role behaviours in order to identify a number of potential anomalies.

System Accessed and Used: Although users may adopt new systems or use remote systems on occasions, they commonly use dedicated systems for work purposes. The identification of user account credentials on another system could indicate that the credentials are being used by an attacker for malicious means.

Date and Time: Activities which fall outside of working days and hours may indicate suspicious behaviour.

Location: User activities which takes place in a variety of locations, during a short time frame may trigger suspicion.

Session Duration: Session duration should correspond with the working day. Deviations may be a sign of malicious activities.

In addition to recording such behaviours, tools are available which can identify how individuals interact with application screens or pages. Through collecting big data and running analytic algorithms, UBA can track anomalies and record the behaviours of users, and may enable the identification and detection of security breaches. In addition to distinguishing between authentic and malicious human users, the developed user profile can also be used to identify actions which may suggest machine-controlled actors such as bots, designed to impersonate humans. Such UBA systems can be incorporated in Information Security Management Systems (ISMS) during 'security by design' planning when frameworks are established to include process rules, security requirements and technology countermeasures. However, the controls that organisations implement are still prone to circumvention, through either malicious or inadvertent actions by humans, resulting in potential leaks or insider attacks. This security challenge has been further complicated by the increased adoption of cloud platforms by organisations, resulting in a shift in the security boundary from technical controls to human behaviour and the effectiveness of identity and access management systems. Proactive monitoring with the intention of detecting malicious behaviour may therefore be one of the most important elements of a security system and may counter the limitation which exists within security tools which assess performance in terms of data loss through exfiltration, as this trigger for reactive actions, rather than proactive mitigations, may not be the most effective solution given that the data has already been lost at the point of measurement. In addition to monitoring exfiltration, effective UBA implementation will provide proactive security by considering activities along the length of the threat actor attack chain; reconnaissance, circumvention, aggregation, and obfuscation, in order to detect and prevent attacks as early as possible.

Data derived through UBA will monitor the movements of users across a network, providing potential indications of reconnaissance activities. Attempts to circumvent installed security tools can be investigated in accordance with policies and rule sets, and can provide warning of any potential malicious interest held by users. Should they successfully overcome security tools, the aggregation of data can provide further evidence, while attempts to obfuscate behaviours suggests a level of premeditation. This can commonly include the renaming of files, which while acceptable practice in many contexts, can be questionable should the file be one that has never been accessed before. As the monitoring and questioning of activities can be burdensome to organisations and will certainly encourage debate about an organisations balance of activities to preserve the confidentiality and integrity of data, while ensuring that it is available to enable and support business processes, the automation of UBA and the use of machine learning may reduce the expensive security analyst resource burden associated with monitoring activities. The development of AI has supported the evolution of monitoring activities, enabling organisations to monitor all entities on an access control list (ACL), privileged and authenticated users, and other entities alike. As a result, UBA has increasingly been referred to as UEBA, which stands for User and Entity Behaviour Analytics. The additional term 'entity' can refer to IT systems, critical infrastructure, processes, organisations and wider nation-states. The behaviour of such entities is analysed in addition to the authorised users, although the entities can take the form of individuals. The nature of the activities is fundamentally similar in using large datasets to model typical and atypical behaviours parties within a network, in order identify behaviours which may indicate a potential threat. The analysis of the data can take place through statistical modelling, rules, machine learning and threat signatures. Machine learning can compare activities to baseline measurements which represent 'normal' behaviours and react in real-time, raising an alert and response request. Such proactive and accurate scanning can enable security analysts to accurately respond to threats quickly before they become exfiltration breaches.

The acknowledgement that preventative and reactive security measures alone cannot provide enough confidence that a network is secure, and that effective proactive monitoring may provide organisations with more time to respond to threats and therefore minimise damage, has resulted in the widespread adoption of behaviour analysis solutions. Despite the identification of these benefits, business concerns attached to the availability of data and operational efficiency will still remain after the implementation of such monitoring if viewed in isolation, so they do not replace the need for a comprehensive ISMS. To support the need to ensure that data is available and to mitigate the risk of users developing ways of overcoming security measures in order to enable them to perform their roles more efficiently, some organisations have increasingly embraced the idea of "Trust But Verify". This view contends that having trusted employees enough to hire them, they should be given access to the systems, applications and physical locations that they need to perform their jobs. In accordance with a system of role- based access control, UBA and UBEA systems can help to verify appropriate access and system use and subsequently detect malicious activities and trigger security responses. Although this appears to be a useful security process, the checks to assess and confirm whether behaviour at an end point has been appropriate, may have some drawbacks related to privacy compliance and the high level of data analysis expertise required to manage large-volumes of data. The monitoring of user activities through collecting any content, key logs, key strokes, or screen shots may potentially breach laws designed to protect personal privacy, therefore requiring for analytics of big data to be carried anonymously by removing information that can be used to identify an individual, and then potentially encrypted to add a further layer of protection. These security measures are required to protect the privacy rights of individuals and should be considered at the onset of a security programme.

6 Privacy by Design

The directed observation and recording of the activities of individuals, with the intention of identifying patterns of daily life has received widespread public criticism in recent years, with such surveillance being viewed to breach the human right to personal and family privacy. Recognising the need to consider the legal rights of individuals and the reputational risks associated with activities which may breach them, technical development operations are increasingly considering privacy protection at an early stage. The concepts of "Privacy by Design" and "Privacy by Default" have been discussed during debates regarding data protection. Privacy by Design is based on the view that data protection is best achieved when it is considered as an integral component of technology design. The effective incorporation of 'Privacy by Design' reduces the need to alter systems later on, often through the laborious deployment of 'patches'. In addition to avoiding future work and inefficiencies, installing privacy controls during the design phase reduces the risk of human error, which may be particularly evident in employees that possess limited IT skills or security awareness. This is also pertinent when services are provided and information shared with third parties and directly with customers.

Technical and organisational measures to ensure security have been incorporated into operational procedures and practices across organisations in the form of data encryption, the implementation of identity and access management tools and through the provision of virtual proxy network tools, but Privacy by Design takes a proactive approach to managing the risk of overprocessing personal data [10, 27]. This approach to security encourages data minimisation actions and the adoption of data protection tools such as anonymisation pseudonyms to take place at an early stage. While Privacy by Design is being increasingly adopted, the requirement to protect data resulting from legislation including the General Data Protection Regulation (2018) has encouraged organisations to review processes and procedures, in order to ensure that they are meeting the obligations, outlined in Articles 25(1) and 25(2) of the GDPR. The UK GDPR presents the expectation that organisations consider the protection of data during all processing activities, including the design phase and throughout the lifecycle of a system, service, product or process. In order to meet the GDPR requirements, organisations are expected to design and implement appropriate technical and organisational measures, in order to support the achievement of data protection principles and integrate appropriate safeguards into processes from the onset of service design. Although the legislation does not clearly stipulate which protective measures should be adopted, in order to encourage the effective adoption of the data protection principles across an organisation, data protection should also be considered during the design and development of organisational policies, processes, and business strategies, as well as being incorporated into development operations,. Through evidencing how privacy has been considered both at an initial phase of a process or system design, as well as being practiced during the course of operations, organisations are well placed to demonstrate compliance with the requirements of GDPR. This should consider implementing controls to ensure that only the necessary data needed to achieve a specific purpose is processed, which may include personal data. This is acceptable when it can be assured that the risks attached to decisions have been effectively assessed and that the individuals concerned have been notified and provided with sufficient controls and an opportunity to exercise their right to refuse for information to be disclosed and destroyed if requested.

The vague nature of the descriptions attached to Privacy by Design leave successful implementation open to interpretation. This vagueness appears to indicate a difference in the understanding between policy makers and technical engineers, of what a technical compliance with data protection should look like. The view that data minimisation can be replaced with technical and policy controls can result in the adoption of security procedures which do not effectively mitigate the privacy risks that are present in big data collection processes. Such databases can contain large amounts of personally identifiable information and sensitive data which are attractive to criminals, competitors and governments. The risk from such threats is exasperated by the challenge of securing big data from insider threat data leaks. The scale of this threat means that a reliance upon the trustworthiness of employees may not be enough to mitigate the risk to data. As data in such stores can be copied, edited and shared, it is difficult to maintain its confidentiality and integrity. Given these challenges, it is crucial that data minimisation is applied from the onset of the design of a system in order to support organisations to meet their privacy obligations and that accurate models are created to support the identification of threats, thereby improving the focus of security operations and associated resource deployment.

7 Threat Modelling

Although differences exist between the definitions given by authors, threats can be viewed as a human action which can influence the likelihood of an activity impacting upon a vulnerability, which can harm an organisation or business process. This can include harm which can be experienced because of a vulnerability within a software application or system process. In attempting to mitigate this, threat modelling is incorporated into the early phases of software design. 'Security by Design' aims to make systems secure by including security measures into the design phases of system design. The identification of all potential threats provides designers with an opportunity to incorporate mitigations from an early phase, thereby reducing the risk of future damage and costs. Threat modelling offers great benefit to security designers by enabling them to prioritise and allocate scarce resources. In addition, the models provide a tool that can be used to increase the level of security awareness within a workforce, through informing employees about threats and the critical network design areas which need to be protected. The identification and classification of threats can be achieved in a number of ways and threat modelling tools exist which consider the impact of a range of attack vectors upon data and processes, at different points on its journey. Threat modelling considers;

(1) The context of an organisation and the environment it operates in
(2) Asset identification
(3) Threat actors, vulnerabilities and the resulting risk
(4) Potential attack vectors
(5) Prioritisation of target hardening opportunities/risk mitigations.

The methods for constructing threat models fundamentally consider a description of an attacker, a description of attacked resources, or a description of software. Classification can be provided through privacy tools, attack trees, attack libraries and processes such as the STRIDE model. The STRIDE model supports Security by Design through considering potential malicious activities within a system [15]. The threats are categorised; Spoofing, Tampering, Repudiation, Information Disclosure, Denial of Service, and Elevation of Privilege. The model considers each security threat at defined points on a Data Flow Diagram and the severity of the impact of each threat upon a business is calculated in order to prioritise the implementation of security solutions. This risk can be assessed through the use of a model such as the DREAD model, which considers threats in terms of; damage potential, reproducibility, exploitability, affected users and discoverability [28]. What is common between the different approaches is that they consider threats based on attack methods and upon the impacts of threats. The classification of threats can group themes together, to provide an overview to guide security strategies, or they can provide a detailed list of attacks and descriptions of the methodology and how they exploit vulnerabilities in a system, which can support the design of security systems. An understanding of both can be valuable, with greater detail benefitting risk-based decisions designed to justify investment in security systems. The defined lists of security threats can be used to justify the design components of an informa-tion security management system, which should aim to preserve the confidentiality, integrity and availability of data through:

(a) Protecting information and its carriers from threats to confidentiality, integrity and availability
(b) Protecting software and hardware and their settings from threats to confiden-tiality and integrity
(c) Defining responsibilities for monitoring the implementation of data security rules.

Given that the unauthorised activities, malicious actions or inadvertent mistakes of users can result in the loss, destruction, distortion or substitution of data, ongoing monitoring is required to identify potential threats and vulnerabilities to information, software and hardware, and to ensure compliance to data security rules. To improve both operational and cost efficiency, monitoring should be as directed as possible and needs to target the "presence of an intelligent adversary bent on breaking the system" [24]. Effective threat modelling can support the deployment of proactive security activities and behaviour analysis through considering potential adversary attacks and the exploitation of network vulnerabilities. Advances in the automation of systems has improved the ability of organisations to identify and analyse the security posture of a system, and the determination of appropriate controls or the defences required to protect assets from potential threat actors and attack vectors. Examples of the successful implementation of automated systems, designed to identify trends and use activity logs with the aim of mitigating potential insider threats have been observed in both large sets of organisational data [8] and in investigations considering the analysis of weekly data, with the aim of detecting small changes in behaviour [25].

8   The Future of Proactive Cyber-Security

Alongside the growing threat from nation state actors and organised criminal gangs, the threat posed by the malicious activities of disgruntled employees has evolved into insiders who offer the use of their legitimate access to assets, into a malicious activities service for sale on the dark web. This threat continues to morph as organisa-tions have to consider alternatives to traditional pre-screening checks and interview processes, and the influence of workplace cultures upon employee behaviour may have changed during the prolonged period of working away from colleague and management oversight due to the global coronavirus pandemic. As a result, organi-sations increasingly contain large number of employees based across the globe who have never met another colleague in person, having interviewed via an online video conference platform. In the presence of a motivation to conduct a malicious action, working away from structured work locations and without physical contact with organisation representatives, may encourage the development of a belief that threat actors won't get caught, should they use their legitimate access to assets and systems inappropriately. In order to counter this threat, behaviour profiling, threat modelling and analysis mechanisms provide an opportunity for organisations to proactively identify anomalies and inconsistent employee behaviours.

With the increased movement of people and devices, the security perimeter has moved from static security tools like firewalls to people, their identities and access. Proactive security has become a necessary component of an effective ISMS, providing time to deploy tools to delay movement across a network, deny access to assets and respond to the threat actors themselves. Behaviour based security systems will enable organisations to better anticipate potential data loss incidents before they occur, thereby potentially removing, rather than reducing harm. This challenge will remain as increasing quantities of data is stored in Cloud environments and becomes reliant upon the security systems of third-party providers. This transfer will not relieve organisations of their duties to protect data, with new and amended data protection regulations arguably requiring for them to assess the risk attached to its use more frequently, potentially including varying degrees of employee surveillance. The incorporation of Privacy by Design, 'Trust but Verify' processes and UBA systems may in fact improve the efficiency of employees by reducing the need for excessive layers of protection, while also protecting personal data. A combination of research into human behaviour, the modelling of threats and user behaviour analytics will also improve the ability of organisations to identify threats and to employ appropriate and effective risk mitigations quickly. The ongoing development of AI will further improve the ability of organisations to proactively identify human and entity threats, while the improvements to AI and machine learning will reduce the number of false positives, thereby improving resource allocation and the cost effectiveness of security systems. Furthermore, improvements in AI systems will increase the ability of security analysts to respond to security threats with greater confidence and speed, provide greater return on investment and visible improvement to organisational resilience to reassure business leaders, and crucially act as an effective deterrent to malicious internal and external actors.

References

1. Allcott H, Gentzkow M (2017) Social media and fake news in the 2016 election. J Econ Perspect 31(2):211–236
2. Appelbaum SH, Shapiro BT, Molson J (2006) Diagnosis and remedies for deviant workplace behaviours. J Am Acad Bus 9(2):14–20
3. Bassett JW (2003) Solving employee theft cases: during investigations, co-workers often provide the best clues as to who is stealing from the company. Intern Auditor 60(6):23–26
4. Broucek V, Turner P (2006) Winning the battles, losing the war? Rethinking methodology for forensic computing research. J Comput Virol 2(1):3–12
5. Brundage M, Avin S, Clark J, Toner H, Eckersley P, Garfinkel B, Dafoe A, Scharre P, Zeitzoff T, Filar B, Anderson H (2018) The malicious use of artificial intelligence: forecasting, prevention, and mitigation. https://arxiv.org/pdf/1802.07228.pdf. Accessed 19 Feb 2021

6. Daigle RJ, Morris PW, Hayes DC (2009) Small businesses: know thy enemy and their methods. CPA J 79(10):30
7. Denver M, Pickett JT, Bushway SD (2018) Criminal records and employment: a survey of experiences and attitudes in the United States. Justice Q 35(4):584–613
8. Gavai G, Sricharan K, Gunning D, Rolleston R, Hanley J, Singhal M (2015) Detecting insider threat from enterprise social and online activity data. In: Proceedings of the 7th ACM CCS international workshop on managing insider security threats, pp 13–20
9. Greenberg J (2002) Who stole the money, and when? Individual and situational determinants of employee theft. Organ Behav Human Decis Process 89(1):985–1004
10. Gürses S, Troncoso C, Diaz C (2011) Engineering privacy by design. Comput Priv Data Prot 14(3):25
11. Heath J (2008) Business ethics and moral motivation: a criminological perspective. J Bus Ethics 83(4):595–614
12. Hedricks CA, Rupayana D, Puchalski L, Robie C (2018) Content of qualitative feedback provided during structured, confidential reference checks. Pers Assess Decis 4(1):4
13. Holzer HJ, Raphael S, Stoll MA (2006) Perceived criminality, criminal background checks, and the racial hiring practices of employers. J Law Econ 49(2):451–480
14. Hunter RF (1984) Validity and utility of alternative predictors of job performance. Psychol Bull 96(1):72–98
15. Hussain S, Kamal A, Ahmad S, Rasool G, Iqbal S (2014) Threat modelling methodologies: a survey. Sci Int (Lahore) 26(4):1607–1609
16. Jahankhani H, Al-Nemrat A (2010) Examination of cybercriminal behaviour. Int J Inf Sci Manag 41–48
17. Kirwan G, Power A (2013) Cybercrime: the psychology of online offenders. Cambridge University Press, Cambridge
18. Kirwan G, Power A (2011) The psychology of cyber-crime, 1st edn. IGI Global, Hershey, USA
19. Kocsis RN (2006) Criminal profiling: principles and practice. Humana Press, Totowa
20. Latto A (2007) Managing risk from within: monitoring employees the right way. Risk Manag 54(4):30
21. Mitre Att&ck (2021) https://attack.mitre.org/. Accessed 19 Feb 2021
22. Parilla PF, Hollinger RC, Clark JP (1988) Organizational control of deviant behaviour: the case of employee theft. Soc Sci Q 69(2):261
23. Pittenger DJ (1993) The utility of the Myers-Briggs type indicator. Rev Educ Res 63(4):467–488
24. Potter B, McGraw G (2004) Software security testing. IEEE Secur Priv 2(5):81–85
25. Rashid T, Agrafiotis I, Nurse JR (2016) A new take on detecting insider threats: exploring the use of hidden Markov models. In: Proceedings of the 8th ACM CCS International workshop on managing insider security threats, pp 47–56
26. Sauser WI Jr (2007) Employee theft: who, how, why, and what can be done. SAM Adv Manag J 72(3):13
27. Schaar P (2010) Privacy by design. Identity Inf Soc 3(2):267–274
28. Singhal A, Banati H (2013) Fuzzy logic approach for threat prioritization in agile security framework using DREAD model. Int J Comput Sci Iss 8:182–191
29. The Information Commissioners Office (2021) The UK GDPR. https://ico.org.uk/for-organi sations/dp-at-the-end-of-the-transition-period/data-protection-now-the-transition-period-has-ended/the-gdpr/. Accessed 19 Feb 2021
30. The Fraud Act (2006) The crown prosecution service. https://www.cps.gov.uk/legal-guidance/fraud-act-2006. Accessed 19 Feb 2021
31. Traub SH (1996) Battling employee crime: a review of corporate strategies and programs. Crime Delinq 42(2):244–256
32. Turvey BE (2014) Criminal profiling. Encycl Clin Psychol 1–6
33. Von Solms B, Von Solms R (2004) The 10 deadly sins of information security management. Comput Secur 23(5):371–376
34. Vroom C, Von Solms R (2004) Towards information security behavioural compliance. Comput Secur 23(3):191–198

35. Warikoo A (2014) Proposed methodology for cyber-criminal profiling. Inf Secur J Glob Perspect 23(4–6):172–178
36. Wells JT (2001) Why employees commit fraud. J Account 191(2):89

Digital Citizens in a Smart City: The Impact and Security Challenges of IoT on citizen's Data Privacy

Robert Benedik and Haider Al-Khateeb

Abstract Cities are investing in smart technologies to improve efficiency, resilience against cyber attacks and quality of life for their residents. They provide enhanced services and can largely benefit the environment too. However, there are a variety of security implications during the transition towards a smart city. For example, anticipated risks have an impact on the citizens' safety, privacy and access to critical national infrastructure. Therefore, smart technologies must be carefully planned to maintain efficiency, security and overall desirability. We follow a holistic approach to review these risks with a proposed CyberSmart framework focusing on four domains namely Cybersecurity risk, Cyber Resilience, Data protection and privacy, and Governance (CRPG). The discussion is formed around data privacy alongside strategic solutions to address related aspects such as the management of operational risk, key stakeholder complexity, and lack of trust in new devices. The proposed framework helps to identify and build mitigating actions to support incident response planning with a focus on prevention and management. It also demonstrates the importance of collaboration to address resilience at a strategic level.

Keywords Cyber natural · Smart environment · Digital citizens · Regulation · User privacy

1 Introduction

The rapid evolution of Information and Communications Technology (ICT) has benefited and restructured how people interact both in their business and private lives. Alongside these innovations, companies developing digital and networked solutions have started integrating three principal technological systems into urban cites: the

R. Benedik
Northumbria University London Campus, Northumbria University, London, UK

H. Al-Khateeb (✉)
School of Mathematics and Computer Science, Wolverhampton Cyber Research Institute (WCRI), University of Wolverhampton, Wolverhampton, UK
e-mail: H.Al-Khateeb@wlv.ac.uk

Internet of Things (IoT), cloud computing and big data analytics. With these system integrations, the "smart city" has been defined. As smart cities are driven by the data input of their citizens, challenges are posed by expectations of privacy and security of systems and data as a whole. According to studies [1], this involves capturing Personally Identifiable Information (PII) as well as data about citizens households and linking this data together to profile citizens to make decisions about them. For instance, research on user profiling shows that people can be classified and identified based on a short segment of driving data [2]. From a legal perspective, where clarity can often be found, privacy breaches are typically covered under laws such as the General Data Protection Regulation (GDPR) in Europe, and by sets of privacy laws across the United States (US). However, these statutes tend to lag contemporary technology deployments, including the quickly evolving "smart city". Besides privacy concerns, there are other many security considerations to be made in deployments of smart cities. Smart cities are prime targets for cyberattacks and the protection of collected sensitive data is of primary concern [3]. It has become evident over time that many manufacturers of "smart city systems" have primarily concentrated on developing functionality at the expense of securing their systems. It also remains unclear who is accountable to fix deployments within a smart city if a system is exploited or crashes, or who defines policy on how a system should be defended against possible cyberattack. It is evident with the development of smart technologies that new risks are not investigated thoroughly and there is a lack of regular forums that investigate this [3].

The widespread use of the Internet of Things which is the foundation of smart cities has caused many questions around security, data privacy and data protection. Therefore, having a robust framework and sophisticated protection models is critical in supporting both academic and industrial areas. Motivated by these reasons, the CyberSmart CRPG framework is proposed for smart city managers and stakeholders to support the development roadmap during the transition from an urban city to a smart city. It is predicted that now and, in the future, mitigating, controlling and managing existing and new risks will be a critical task due to the data and security risks and impacts highlighted in this paper.

In the remainder of this paper, Sect. 2 discusses the characteristic of smart cities from the literature review point of view of similar works in the field, followed by the current unidentified risks and vulnerabilities in smart cities in Sect. 3. Section 4 proposes a framework on how the different demands and aspects can be governed. Section 5 details a roadmap on the deployment of the framework against a model smart city. The conclusion and statement of future work are then shared in Sect. 6.

2 Background and Related Work

Increasing populations within cities mean that managing growth is a rising concern. With this sustained growth, urban areas are facing challenges and pressures that this influx brings. By employing more sophisticated technology, concerns such as traffic congestion, public safety and sanitation can be addressed and managed more efficiently and effectively. The building blocks of a smart city support these developments, incorporating networks and communication, processes, data management and security, trust and privacy [4]. In return using these technologies in a smart city may result in emerging and increased exposure to risk, which needs to be catered for properly.

2.1 Building Blocks of a Smart City

Urban agglomerations can trigger the development of a variety of problems that need solutions. The Smart City concept addresses this and has already demonstrably improved many factors of urban life. Examples include how transportation is monitored and used, supporting the education system, monitoring energy usage and allowing individuals to monitor their health and medical treatments [5]. These challenges are addressed through creativity, cognition, and cooperation among relevant stakeholders to design smart solutions [6].

Benefits from smart solutions have been recognised by governments, both local and national, and as adoption rises it begins to improve living standards and drive the move towards big data applications. It is recognised that many sectors of a city's economy can be enhanced by big data applications. An example is surveillance monitoring through the usage of devices such as CCTV, which can be used to relieve congestion dynamically or protect the public by monitoring, detecting and preventing crime in its early stages [7].

At the outset of this burgeoning rise of smart city technology, there were a variety of different terminologies used to describe its constituent parts, including digital city, hybrid city, ubiquitous city, information city, wireless city and intelligent city [8]. The concept of a wireless city was often extremely confusing as people linked this to wireless communication such as public Wi-Fi provision. It became apparent that the varied names caused confusion, and the term smart city became the preferred choice. The word "smart" itself suggests artificial intelligence and automatic technologies such as self-configuration, self-healing, self-protection, and self-optimisation are used extensively across the components of the model [8]. As a whole the term smart city signposts technologies that allow modern cities to grow and thrive through increased productivity using clever solutions. Overall, smart cities use technology, to automate and improve city services [9], they are built from a network of items, these items are embedded with sensors connected to the Internet.

In a research paper on the concepts of smart cities, Sikora-Fernandez and Stawasz discuss that cities can be defined as smart if they have the dimensions [10] illustrated in Fig. 1.

To understand a smart city, it is important to realize how the technology supports its further development and how it supports new opportunities to promote a more sustainable and resilient way of living. A variety of studies have been conducted to research the components and infrastructure of smart cities; some have focused on

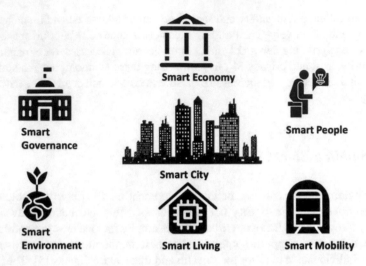

Fig. 1 Smart city

the international point of view on smart buildings [11] and in favour of smart cities having a positive impact on the environment if used to their full capability. They also recognized improvements possible in security management using the technology provided, and the low operational costs to deliver the benefits [12, 13]. Conducted a study where the author evaluated the efficiency of every energy source to investigate how smart energy systems supported a sustainable future. The authors concluded that, if products from the same energy source increasing then a greater efficiency is achieved with lower emission levels. During the study, Dincer and Acar were in favour of geothermal energy due to it being a cleaner and more sustainable resource. [14] Carried out their research on the Internet of Things (IoT), and what the impact of smart water management would be on businesses. The creators proposed a model that would support both urban cities and extend beyond that to rural areas, which would be feasible if IoT development was integrated into smart city models. In [15], the authors researched smart mobility and transportation which highlighted the rapid expansion of Information and Communication Technologies (ICT) in smart cities. The authors suggested new frameworks should be created where people should know what they want and what they need to bridge the policy gap. Without this careful definition then policy implementation would fail.

The building blocks of a smart city can be categorised into four as shown in Fig. 2, each of which will be discussed in the following sections [6].

Fig. 2 Building blocks in a smart city

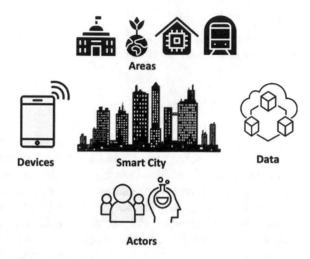

2.2 Actors in the Smart City

To ensure a smart city is developed successfully, it is critical that there is an understanding of the needs and wants of the actors (stakeholders) within a city. Therefore, there is a requirement that time and research is invested to understand the market and what the key drivers and motivations are to appeal to the majority. The final solution of what a smart city looks like may be similar each time, but by understanding the actors it can be implemented by really focusing on what their motivations and interests are to be most effective. Actors within a smart city fall into the criteria below [6]:

- Individuals (residents, visitors, city activists)
- Businesses
- Vendors (hardware, software and system integrators)
- Government (national, regional, local)
- Academics (public, private or independent researchers)
- Organizations (multinational, non-governmental, philanthropies).

2.3 Areas of Application of Smart City Technologies

When identifying areas within a city that can be improved by smart city technology, some are more receptive than others to enhanced technology. The main areas of focus are utilities, mobility, safety, health and education.

- Utilities (smart grids, smart meters)
- Mobility (smart parking services, smart transportation, smart traffic management)
- Safety (smart surveillance, smart identification)

- Health (smart healthcare)
- Education (smart education)
- Governance (smart governance)

2.4   Devices (What is IoT?)

Devices are recognised as different things to people. These could be smartphones, laptops or smartwatches for example. However, when looking into the building blocks of a smart city, the definition of a device expands way beyond this. Devices are used at a much larger scale and often connect and communicate with one another (machine-to-machine—M2M). The Internet of Things (IoT) is what makes the connection of these devices possible. IoT has different interpretations for different people. Initially, the classic internet was designed to be a network of computing devices and networking equipment (servers, routers switches firewalls etc.). However, IP cameras, various sensors, and smart assistants, smart meter readings, smart home appliances (such as Alexa), smart locks are some examples of when a physical object is developed with connectivity, computing, sensors and actuators to become an IoT. It should be noted however that many of these devices may not be attached to the Internet at all, but to private or encrypted networks to maintain their integrity. The term IoT may continue to be applied to these devices/sensors. In Fig. 3 we can see many of the possible components of an IoT.

But why IoT? By combining multiple inputs from the physical smart things, the overall outcome can increase in value for the user. The use of smart devices provides the opportunity for monitoring and analysis and to close the system loop—in turn allowing the production of an intelligent outcome. The beneficiaries of IoT include

Fig. 3 The IoT

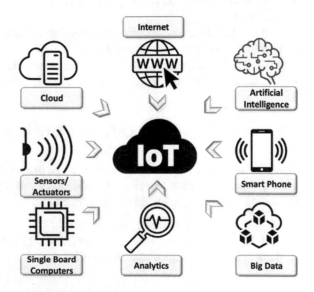

Fig. 4 IoT architecture

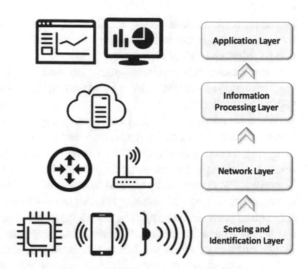

end-users as they gain the ability to monitor and control devices remotely, thus in turn they can use "the things" efficiently by saving time and resources. In turn, the lifespan of the service can be extended, and overall experience improved through this collaboration. Likewise, manufacturers benefit by linking IoT to smart devices to develop the value of the services around the product. By monitoring devices' in real-time, their location, condition, usage and performance can be monitored and tracked. This assists in product improvement, to offer enhanced services and upgrades in the lifecycle. Aggregating data too from multiple sources gives valuable insights which can then be monetised potentially, depending on user agreements/anonymization etc.

IoT is generally categorised into the following categories:

- Consumer IoT (smart home devices)
- Industrial IoT (agriculture, wind farms)
- Civic IoT (smart public transportation, smart water supply, smart electric grid).

There is no specific set architecture for IoT, but we can use the IoT architecture in Fig. 4 as a guideline. Within the architecture is a base layer that manages the physical interaction through smart sensors to measure, feel, collect and control instruments. This data is then transferred into the upper layer, which collects the information and determines how the information is routed to alternative networks, devices and IoT.

2.5 Data

The new most valuable resource in the world today has changed from oil to data. A modern city runs on data. In a smart world, data is a critical asset of an individual as well because it contains a vast amount of valuable personal and sensitive information including bank account details and history, medical records [16], passwords and location coordinates. This data can be accessed if the hardware or software is hacked due to vulnerabilities in its security, or via social engineering of someone who has legitimate access. It should be assumed that threat actors are actively targeting devices and that new vulnerabilities may be identified, spread and be used criminally—so vigilance is called for.

When it comes to gathering and processing large amounts of data, there are a variety of ways this can be performed, including capturing, transferring, storing, analysing, visualizing, securing and ensuring privacy [17]. Data set sizes vary, but over time are expanding as computing analysis allows and range from Gigabytes to Terabytes, Petabytes, Exabytes, and Zettabytes. Data and data management can be extremely confusing, and creating policies of how to access it, protect it and interpret it can often prove challenging. There are a variety of aspects linked to data itself, such as regulations and ownership which also need deep consideration by data processors. How a smart city can use the data it collects is reliant on many factors.

Cyber-attacks are an increasing threat due to the value of data and the level of information it contains. Consumers and companies often find that they have little control over data as IoT security does not alert them when data is being misused. This leads to an increase in regulatory risk and a requirement to strictly maintain data privacy.

Security is a critical component for computer systems. However, the level of security can largely differ across a range of IoT systems. Many of these issues can be due to inefficient security features, security design errors and default passwords as an example. IoT nodes may have a lifetime of several years, therefore, if there is a security issue on a node, it will affect the whole IoT system for the remainder of its deployment. Security also encompasses privacy, protecting the theft of data and defining access permissions between data of different sensitivities within a network.

3 Preparing for Unidentified New Risks and Vulnerabilities Within a Smart City

Many risks and vulnerabilities are identified and expected in a smart city. The array of devices, technologies and open systems that are interconnected does not come risk-free. While the expected risks and vulnerabilities are managed with controls and contingencies implemented, there are also those unexpected vulnerabilities that need to be managed. This discussion delves into the challenges that are faced by smart technologies and how smart city managers can mitigate and manage them [18]. Figure 5 shows a high-level Strength, Weaknesses, Opportunities and Threats (SWOT) analysis of factors influencing Smart Cities.

Three key technology attributes to support the creation and reasons for having a smart city, but in turn, they also carry their risks are:

Strengths	Weaknesses
Improvement in transportation, both in time efficiency and alternative methods, increasing convenience	Increase in security and data breaches impacting individuals and businesses both financially and reputationally
Eco-friendly technologies and ways is where the public will be educated on how to become more eco-friendly	Access to funding and ability to use available funding to effectively manage collaborative investments.
A smart city becomes a better place to live and attracts more residents thus increases revenue	Lack of expertise in specific fields such as security, risk management and overall up to date knowledge
Improvement in public services such as emergency services and security services improves public healthcare and safety	Citizen buy-in and general understanding of smart cities proves challenging due to education
Opportunities	**Threats**
Creation of opportunities for individuals and businesses in both career opportunities and business expansion	The development of transformation can also threaten the environmental impact of traffic congestion
Opportunity to collaborate with a variety of experts to develop proposals and access funding and revenue streams	Individuals health with further reliance on transform and smart devices to perform tasks that once required physical excursion
The opportunity to share best practises and collaborate with other smart cities to build a stronger framework	Increased stress on services with a growing population with a more critical impact if systems were to fail
Opportunity of key stakeholders within the smart city to prioritise areas which need further investment to improve development	The risk of the government introducing spending cuts resulting in the inability to roll of technologies effectively

Fig. 5 SWOT analysis of factors influencing smart cities

1. **Speed and Range**: Interconnecting smart devices prove extremely efficient in daily lives.

2. **Interconnection**: Complex interdependencies across services are created. Networks support critical services.

3. **Innovation**: The use of innovative technologies provides the novelty of many enhanced services.

While these all show how a smart city can provide many advantages there are also increased risks;

- Operational Risk: Due to the increased reliance on technologies and devices, significant impacts can be incurred due to inadequate or failed processes and systems.
- Management Complexity: Larger areas of expertise are required, and agreements may be harder to reach.
- Ambiguity and Lack of Trust: The level of data that is stored is a growing concern especially with the media awareness of incidents.

3.1   Challenge #1: Operational Risk Management Within a Smart City

The threat of increased operational risk within a smart city is largely down to the scale and complexity of interconnected technologies and devices [19]. Not all cybersecurity managers are aware of the scale of vulnerabilities that are linked to devices that are spread across a city [20].

Smart technology is operated in silos and often split across a city's operational domains [21]. These are integrated to support both critical and non-critical processes. Organisational silos can limit the advantages of smart technology and due to the large interconnections can significantly increase the operational risk. Due to this interconnection, if one device contains a failure this is likely to passes on to connecting devices and potentially creating major disruption. Key stakeholders within IT and cybersecurity are required to manage the risk across this broad structure of interconnected devices and technologies and minimise the risks and manage incidents along with security managers.

Operational risk becomes more apparent due to the fact there is so much to learn within how smart city technologies operate and there is a shortage of qualified Smart City and cybersecurity personnel. AI technologies open up new risks as these are more unpredictable in their decision making and data handling. Cyber-attacks may also be incurred from members within the framework who identify a vulnerability and use it to advantage such as stealing data for financial benefits. The more dependant the technology becomes, the more of a target it can be.

3.2   Challenge #2: Management Structure and Key Stakeholder Complexity

With a smart city comes the management and stakeholder complexity. Due to existing processes and services already being available a smart city builds additional layers, which means that processes can range from basic manual ones to brand new automated smart technology ones.

This interconnection and complexity create more opportunities to identify weak security points and take advantage of this. Due to the array of risks that would be present, it is difficult to allocate a risk owner within the management structure. Furthermore, the newness and advantages that may be seen in certain devices may overpower the ability to identify risks. This also comes with the limited expertise within cybersecurity in smart city officials. Further development needs to be completed on risk management and framework standards. Another added complication is the ability for stakeholders to communicate what smart city device requirements are needed accurately to external stakeholders. Lack of authority means that a city could invest in technologies that can be damaging to the city and its citizens.

3.3   Challenge #3: Lack of Confidence and Trust in New Devices and Technologies

Distrust and uncertainty are other major challenges that come with the development of technology. For example, when a major data breach has occurred from a fault in a newly installed system, this has a major impact on the public's confidence often resulting in people not wanting to use a potential service [19].

As much as there are some clear benefits for certain technologies which collect data, for example, security surveillance, having so many interconnected devices cause unpredictable behaviour [22]. If the behaviour is unpredictable it means the risk is more difficult to manage and it can be likely that inadequate controls are in place.

4   CyberSmart CRPG Framework

There are several factors for consideration when developing a smart city. Many of the aspects that form a smart city are integrated over a long period and piece by piece. There are also different reasons as to why a city chooses to develop into a smart city. This could be to become more environmentally friendly or to rebuild after a catastrophe or even to improve general day to day living and convenience. Due to the nature of smart cities, new or enhanced risks are a growing concern. In the design stage of smart city technologies, these risks and related controls need to be assessed. Figure 6 presents the different stages within the CRPG framework and their related activities and objectives.

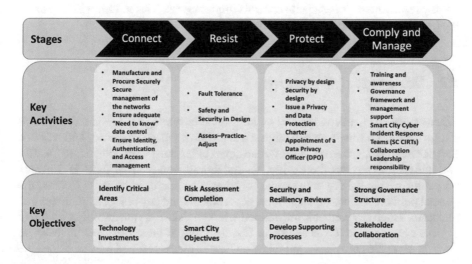

Fig. 6 Smart city—CyberSmart CRPG framework

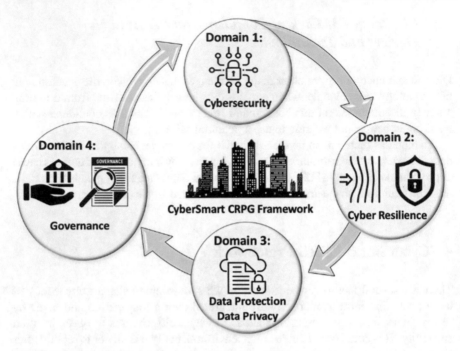

Fig. 7 CyberSmart CRPG framework

In Fig. 7, we demonstrate the 4 domains addressed by our framework to facilitate such assessment.

Additionally, due to the extensive range of cybersecurity and data risks that exist within a smart city, Fig. 8 shows elements proposed to ensure these risks are thoroughly researched with adequate controls implemented. This in turn will maximise the benefits that a smart city has to offer.

4.1 Domain 1: Cybersecurity

When assessing privacy and security within smart technologies, consideration of risk mitigation must be made for both digital and physical aspects of privacy and security. People who operate these systems within smart cities must work together to identify and mitigate these risks thoroughly. To collaborate and manage these endeavours all parties within the city must come together. This includes business owners, the general public, religious organisations, governments and any other communities and residents. Three key elements are defined within a risk (R): vulnerability (V), threat (T) and consequence (C) as shown in Fig. 9. If any of these elements increase, so does the risk. This provides a guideline on how to assess risk and how advanced the

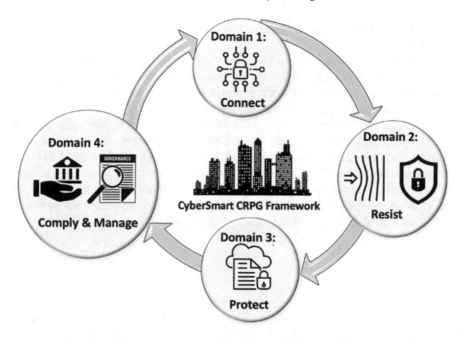

Fig. 8 Connect, resist, protect, comply and manage elements within CyberSmart CRPG framework

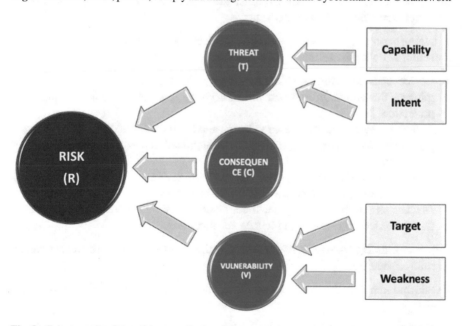

Fig. 9 Cybersecurity risk and its contributing factors

controls need to be to mitigate the risk. This is presented in a mathematical formula: $R = V \times T \times C$.

This same risk expression is also relevant within cybersecurity and privacy. Within a smart city and its connectivity, thorough investigations are required for cybersecurity and privacy risks within all fields of a smart city environment. When identifying vulnerabilities and threats within a smart city, enterprise information technology (IT) environments can be used as model guidelines as they contain similar cybersecurity risks and threats. Having said that; smart cities contain a much wider range of technologies and implementations, so there are likely to be more risks identified. In many instances, it has been evident that security and privacy risks are taken underestimated. Cyber-defence is required to be managed effectively by monitoring physical cyber interactions and producing effective policies that communicate effects and systemic risk. This should all be embedded within planning and operations.

4.1.1 Connect

Due to the extensive range of cybersecurity and data risks that exist within a smart city, the general philosophy "Connect" has been developed and proposed to act as a framework to ensure these risks are thoroughly researched with adequate controls implemented. This in turn will maximise the benefits that a smart city has to offer. The approach comprises four elements:

1. Manufacture and procure securely: Devices should be securely manufactured and enforced with the following characteristics:

- Devices must have the capability to allow regular maintenance checks which will mitigate the vulnerabilities and utilise cryptographic integrity and authenticity protection
- Default passwords must be able to be changed to user-defined and user-managed passwords. By implementing two-factor authentication will create greater security protection
- Cryptographic functions must be utilised with devices and systems, encrypting communications sent from devices must be a necessity
- Devices should be correctly certified. Conformance tests should be completed that are based on recognized security certifications

2. Secure management of the networks: Use the comprehensive zero-trust approach to enforce security.

Create and maintain an asset list of devices and applications to minimise the threat landscape.

- Deploy network segmentation for better access control and improved security
- Integrate logs to security information and event management system (SIEM)
- Take the whitelisting approach (whitelist only legitimate applications) and block everything else

3. Ensure adequate "Need to Know" data control: Provide education on data obtaining and storage to ensure that only necessary data is collected, stored and maintained correctly.

4. Ensure Identity, Authentication and Access management: Ensure that every device is identified and authenticated on the network supported by strong authentication methods such as Multi-Factor Authentication (MFA). Cloud-enabled authentication functions can be used to support internet-connected devices.

4.2 Domain 2: Cyber Resilience

Within smart cities, cyber resilience is in place to ensure critical outcomes and processes are delivered in the event of a major disruption and low impact everyday glitches. Due to the reliance smart cities have on technologies, it can cause major disruption if there is a system failure or an outage on its key technologies. Therefore, cyber resilience is a crucial element as it will ensure that a contingency is in place to adapt and recover in the event of a technology failure [23]. Smart cities are required to provide services and technologies to their citizens while also protecting their safety [24] and security [3]. This also includes monitoring daily operations and vital signs such as emergency response times, pollution risks or traffic control. Cyber resilience requires adequate risk management to identify and prioritise these elements as it covers such a wide specification. In the event of a cyber risk coming to fruition, there is a wide range of impacts that can occur that resilience measures and actions must be taken across regions and nations, including third-party managed network and cloud computing infrastructures.

4.2.1 Resist

The term "Resist" is a design feature or contingency practice in the case of a failure occurring. The objective is to ensure that any attack surface is minimised to provide resiliency and robustness. Contingencies will be made available to provide an alternative solution whilst incident management practice will manage communication and response. There are a variety of methods by which cyber resilience can be maintained and managed. Under "Resist" three high-level possibilities of actions are suggested:

1. Fault Tolerance: When creating adequate contingencies, it is important to first identify operations within the smart city where if operations were to become unavailable it would have a higher or critical impact.

- **Infrastructure Redundancy**: Infrastructure redundancy for critical networks components is vital for a Smart City being operational 24 × 7. However, having the additional capacity for resilience comes at a cost.

- **Data backups and recovery**: To minimise the risk of data being lost, it is essential that the data is stored in several locations, and with at least one off-site copy. This provides a backup service to ensure the maintenance of the running of the infrastructure. This may also include a workaround where an alternative device is used for a temporary period to maintain the process.

2. Safety and Security in Design: Devices and systems should be embedded with alternative contingencies in place. If a device becomes unavailable due to a technological fault or planned maintenance where it needs to be out of action, the contingency method can support this service for a temporary period.

3. Assess–Practice-Adjust: Ensuring resilience in a Smart City requires constant preparedness should any kind of scenario occur. This should include simulated outages, stress testing and penetration testing of the infrastructure. Practising these scenarios can improve the business processes which are already in place based on the outcome of these scenarios. The Assess-Practice-Adjust cycle must occur frequently within Smart City operations.

4.3 Domain 3: Data Protection and Data Privacy

A sound understanding of data privacy and the impacts of materialising risks are critical [25]. An enormous amount of data is collected and analysed daily by using data mining and statistical methods. These methods may result in limited data security, leading to potential disclosure of personal and sensitive data which could have an extremely detrimental impact.

Due to the constant learning and understanding around what privacy is, it has historically been extremely difficult to define the term privacy, especially since privacy definitions have been reflections on contemporary contexts. In 1967 a definition was released by the International Commission of Jurists in Stockholm, which captures the importance of privacy: *"The right to privacy is the right to be left alone to live one's own life with the minimum degree of interference"* [26]. Previous to that the Universal Declaration of Human Rights released earlier in 1948: *"No one shall be subjected to arbitrary interference with his privacy, family, home, or correspondence"* [27]. One of the main concerns contemporarily is around the subject of big data and privacy. When obtaining the materials required for collation and analysis, the risk of privacy is always a concern as care must be taken to ensure the appropriate level of data anonymisation is used. However, it is extremely difficult to maintain anonymity [28] whilst collating large quantities of data which can potentially result in correlations in the data allowing the users to be uniquely fingerprinted.

4.3.1 Fair Information Practice Principles

In 1974, the US Congress approved the Fair Information Practice Principles Act. This discusses how data is collected and that there must be a rationale as to why it is needed. It covers the importance of transparency and giving notice to people on when, and why, the data is being collected. These principles may be relating to now out of date reports, but cover the fundamentals of national privacy laws and policies in various countries:

- General Data Protection Regulation (GDPR)—European Union
- Personal Information Protection and Electronic Documents (PIPEDA)—Canada
- Californian Consumer Privacy Act (CCPA)—United States
- OECD Privacy Guidelines (The Organisation for Economic Co-operation and Development—International standard

In the taxonomy of privacy, privacy could be jeopardized and breached in many different ways [29]. The below four domains were further split into subcategories:

- Information collection: surveillance, interrogation, recording, monitoring people's conversations and activities
- Information processing: processing information that was collected without the subject's consent. Not being transparent on what the data held on the subject is or providing the opportunity to correct this if the data is inaccurate.
- Information dissemination: breach of confidentiality, disclosure, increased accessibility, blackmail, appropriation, distortion
- Invasion: intrusion, decisional interference

It is important to understand what data can be exposed voluntarily and the impact this then has on its protection. People have different boundaries on what data they wish to protect. Social Media is a huge example of this, some people regularly post their activities where profiles on people can be developed on their routines, hobbies and interests. Others feel the need to post little, or not at all, and are extremely cautious on what they allow to be made public.

The overall problem with big data is that its content cannot be wholly erased. Just as libraries used to keep old books in printed format, the Wayback Machine (https://archive.org/) is something like a search engine that archives pages on the web. This data is intended to be retained forever as webpages themselves are ephemeral. This means if there is something stored here that someone legitimately no longer wishes to be available to the public under the EU's GDPR laws, failure to remove it can be interpreted as a privacy violation. Copies from the Wayback Machine can also be made in an uncontrolled fashion, meaning that there is no way of knowing how many copies of this data are stored, or where. Depending on the content of the data it could have severe consequences to somebody's wellbeing, and the location of that data is important to its removal. There is no compulsion to remove data for example stored outside the EU GDPR area in response to a requirement—so it may exist perpetually. As a compromise to make data extremely hard to find, people work with intermediaries such as search engines which are often the portal to this data. If the results in the search engines are masked, then this makes it extremely difficult for the data to be obtained from the main search engines in a territory-bound manner.

4.3.2 Breach of Privacy

The historical influence and development of technology privacy laws show a continuous focus on privacy concerns. The law has often failed to keep up to speed with technological development, which is why broad principles like FIPPs are important. When there has been a development of an innovative new product, such as the smartphone, where all security and privacy concerns are taken into account before launch? If inventors launch a product or service today and have not taken into account the in-depth privacy laws, who will be held accountable and potentially face significant fines and in some cases prison sentences, should privacy of the consumer be breached? Trivial examples of data breaches could be from leaving a laptop on the train, dropping a smartphone in a bar or clicking on a suspicious "phishing email" which results in the user entering data into a fake or hacked website.

When it comes to a personal data breach, not only is it that the data has been accessed that is the concern, but also what is done with that data. Ransomware is one example, where the threat to publish or pass on a victim's data can be used. If a data breach incurs this data can be sold, corrupted, deleted, lost or destroyed. These can have serious and devastating impacts on businesses but most affected individuals. If a security breach occurs, the type of breach must be identified promptly. If it has turned into a material personal data breach, some actions are required to be taken to monitor, control and report this. In the UK the Information Commissioner's Office (ICO) may need to be informed as part of this process. If a company fails to adhere to the personal data breach reporting and monitoring processes, it could be responsible for failing to protect customers from an ethical and practical point. Resulting in not managing data breaches effectively from a practical point of view can lead to significant government fines and potential lawsuits leading to prison sentences. When new technology is developed privacy must be always at the forefront during the development stages. In the same breath, it is also essential that companies are using the latest tools available and their technology is being maintained to an adequate level to prevent these devastating consequences.

4.3.3 The Current Approach to Privacy Breaches/Harms

The European Union privacy laws are embodied within General Data Protection Regulation (GDPR). The US has a variety of privacy laws that are mainly "state-specific" with California perhaps being one of the strongest sets. Within both the US and EU individual personal rights, which include the generation, use and disclosure of personal data along with the responsibilities of the data controller are covered within the practical principles of FIPPs. There are many common factors between the US and EU when it comes to privacy harms and implementing privacy protection within the design of technology, with the differences mainly occurring in implementations, such as obtaining content and notification of breaches. EU legislation is standard across all domains and applies equally to all data controllers. The US however does not implement this transparency. In the US privacy laws and legislation are domain-specific. There are also differences in how privacy by design is promoted, how enhanced data security is implemented and how access rights and data consent is applied. Due to the differences in how legislation and policies are implemented, it can cause some confusion and result in increased compliance and regulatory risk. Smart city technologies are constantly challenging this legislation and policies with regards to privacy.

4.3.4 Smart City Security Concerns

When it comes to smart cities, there are two main security concerns:

1. Security of physical ICT prone to cyber attacks
2. Security of data at rest, in use and motion across the technologies and infrastructure

For a smart city to operate in the way it is intended, it utilises a complex build of groups of digital technology and ICT infrastructure. Due to the heavy reliance on software, there is always the risk of cyber-attacks and new vectors being discovered. Depending on how a device is networked can result in a variety of different entry points from where a hacker can launch a cyber-attack and attempt to alter, disrupt, destroy, deceive or degrade the systems and data.

Cyber-attacks fall under the below three forms focusing on the CIA triad namely Confidentiality, Integrity, Availability:

- **Confidentiality attacks**: This will look at data that is held on the device, how to extract the data and monitor activity to potentially obtain further confidential data (packet sniffing, password attacks, port scanning, phishing, social engineering)
- **Integrity attacks**: This will look to manipulate/change its information and settings; this could also involve installing malware and viruses [30] (Data diddling attacks, Man-in-the-middle attacks, Session hijacking attacks)
- **Availability attacks**: These focus on obtaining full control of the system to force them to be closed or restrict/decline service (DoS, DDoS, SYN flood/ICMP flood attacks, electrical power attacks).

For smart cities to mitigate the risk of the attacks such as the above, its overall security model needs to be comprehensive, later and in-depth, to be cyber resilient and explicitly incorporate the city's privacy and data protection responses.

4.3.5   Identifying Data Privacy and Security Risks Within a Smart City

There are a variety of different concerns linked to smart city technologies and data privacy. One aspect to review when assessing the risks is to look at the ethical consequences of mass surveillance (intended or a side-effect) that comes with smart city technology. This includes predictive profiling of habits and social sorting for example through the use of big data analysis [31]. There are often pros and cons relating to these points when implementing new technologies. If mass surveillance is used as an example, there is an argument that the surveillance is a breach of people's values and basic privacy rights. On the other hand, mass surveillance can be used to support businesses by monitoring footfall, product demand and trending behaviours, thus creating commercial opportunities and jobs through expansion and growth by reacting accurately to market needs [31].

There is no black and white when it comes to privacy. Through the constant innovations and new technologies implemented within smart cities that touch on our daily lives, it no longer becomes a question if an individual would like to remain completely anonymous and not disclose any personal information. It would now be extremely difficult to live an anonymous life, for example avoiding online purchases, using email, smartphones and online banking which are almost ubiquitous in the modern world. Privacy however is still protected by legislative and regulatory measures and remains a significant focus for many individuals. The aim is to ensure the development of new technology incorporates the implementation of proportionate and regulatory-defined data privacy and security measures. There is no one definitive answer or implementation. A suite of controls needs to be reviewed and implemented to ensure that risk is mitigated or controlled. These controls focus on a range of aspects such as regulatory and legal requirements, technical security, and data storage to name a few. Risks and controls will always need to be regularly reviewed as new technology brings new opportunity and risk, and a cyber-attacker can discover new methods to target a network to penetrate or compromise it.

Moreover, ethical concerns play a large role in smart privacy, for example how mass surveillance is managed. Mass surveillance, such as CCTV has a range of advantages and disadvantages, for example, it can be considered extremely useful in protecting the public if a crime has taken place to understand what happened and assist in identifying the suspect and stop it from happening again. Another positive factor would be promoting business growth and reaching out to a target audience by monitoring trending behaviour linked to certain products, which in turn supports business expansion and the economy by providing job opportunities. An argument against mass surveillance is that individuals may not always be able to consent to personal data being held on them and they are uncomfortable with how the data is being managed and used beyond their explicit permission. To live in the modern world, the general use of email, social media and online shopping all fall under the mass surveillance category—under general terms and conditions—therefore it is extremely difficult for an individual to take control. It is therefore essential that data privacy regulations and security measures are considered when implementing smart technologies. A suite of controls is used to ensure regulatory and legal requirements are adhered to and are required to be reviewed regularly to identify any new risks that could potentially arise.

Another issue with this is the understanding of terms and conditions. Because they are often not fully understood and can be extremely complex, people cannot always act in their own best interests when it comes to protecting their data and privacy. A second market solution is companies promoting the advantages of consumer privacy and data security and developing this into their technologies to attract consumers and develop consumer loyalty. This approach as it is unable to garner revenue from behavioural tracking characteristics is likely to be offered with a fee and will ensure that privacy and security settings within the technologies used are regularly enhanced to ensure optimum protection for customers. It will also allow individuals to control and manage their privacy granularly along with supporting companies and public authorities to enhance the protection of operational security, data resources and customer privacy.

Privacy Enhancing Technologies (PET), as defined by the European Commission, are also used, which support the protection of Personally Identifiable Information (PII) and controls how PII should be managed by "different services." PETs help protect PII on websites and smart devices, they are also embodied in ad blockers, cookie blockers and removers. Smart cities, also manage data that is handled by data brokers, this data can be obtained by surveillance devices and card readers as an example. Other alternatives to PETs intended to protect confidentiality are private information retrieval (PIR), statistical disclosure control (SDC) and privacy-preserving data mining (PPDM).

4.3.6 Policies, Regulations and Legalities

Regulatory and legal requirements need to be under constant review to keep up with the constant technological development. When it comes to "urban big data" privacy and security concerns need to be thoroughly addressed. The policy, regulatory and legal landscape needs to be revised not only within smart cities but also at a national level through active revision to ensure it remains fit-for-purpose.

4.3.7 Fair Information Practice Principles (FIPPs)

FIPPs clarifies the main principles when it comes to data use, handling and storage along with the responsibilities of the data controllers. When FIPPs is put into practice there are some significant difficulties. Due to the importance of FIPPs and growing concern of data use and abuse, it has become apparent these controls are essential, and thus many countries have published their revised sets of FIPPs. In the US it was via the Consumer Privacy Bill of Rights in the EU—the General Data Protection Regulation (GDPR).

It is argued that there are gaps in FIPPs, one example is that there is still a huge risk on how data can be shared and manipulated which is not always explicitly called out and if it is not explicitly prohibited will be exploited.

4.3.8 Privacy by Design (PbD)

FIPPs would play an important role in supporting the development of smart cities through the adoption of privacy by design. As introduced and expanded by Ann Cavoukian, the key principle linked to Privacy by Design (PbD) is that privacy is the automatic blanket mode of operation. Therefore unless an individual specifically agrees to data being made available or used, it will automatically be processed and stored as private. Privacy is embedded into key principles which allow for different modes of implementation.

This solution is extremely advantageous to the individual and their control, positioned as positive-sum rather than zero-sum where the aim is to maximise privacy rights and security rather than having to look at compromising these by relying on trade-offs.

4.3.9 Security by Design (SbD)

Security by design works alongside privacy by design. Risk assessments are completed and revisited in the early stages of a product launch to act as a preventative measure in providing security protection. This process includes in-depth testing and pilot testing within a laboratory environment to ensure security measures are effective and comprehensive before a product is fully launched. Cybersecurity plays a major role in the whole process and the product needs to be monitored throughout its lifespan to ensure cybersecurity measures remain effective with necessary action taken if a fix is required or if a data subject needs to be made aware of a potential breach or risk [32]. Security by design encourages users to also take preventative measures, such as using strong access controls, password management and end to end encryption.

4.3.10 Protect

Due to the nature of smart cities, there must be robust privacy and data controls in place as the risks are much higher due to the level and quantity of data held. Record retention policies should be followed to ensure the retention periods of data are fair, only data required to be held is obtained and the use and processing of data are transparent. To ensure data security is practised, cities have privacy and data protection principles in place front and centre. To help implement these principles, the following are three basic key principles that should be applied and followed in smart cities:

1. Issue a Privacy and Data Protection Charter: This charter will encompass guidelines on implementing privacy by design (PbD) and privacy by default when developing smart city devices. This will not only mitigate risks on security crimes but will also provide the public with the confidence that data management is secure.

2. Promote Transparency and Appoint a Data Privacy Officer: This will ensure someone is responsible for incident management and risk oversight within data protection and privacy. Transparency will be a priority, as regular reporting will be issued to show the strengths and weaknesses within the smart city and any breaches that need to be raised. This person will ensure devices and individuals are compliant with policy and regulations and spread awareness on best practices and be responsible for them.

3. Establish Data Governance Contracts with Third Parties: Third parties need evidence that they are compliant with privacy and data protection regulations. If the third party provides a service where they are managing individuals' data, not only do they need to be compliant with regulations, but contracts need to be in place to ensure clarity on who owns the data and who is responsible for incident management should a breach be incurred.

4.4 Domain 4: Governance

Companies could potentially not take into consideration what the key priorities, needs and interests are of the city and its residents and act solely in the interest of the company [18]. Due to the smart technologies being in such an early stage, cities need to be cautious in the long-term investments that are made. Smart City officials must monitor developments and intervene when necessary for both private and public interests. Governance in this field encompasses two complex factors:

- **Cultural Diversity and Disparate Visions**: Residents within a city consist of a range of different cultures and principles. Not every individual and business will have the same objective, therefore, collaboration across a range of governmental, business, residential and other stakeholders is essential in understanding the needs of the city and how this will fit into the smart city objectives.
- **Fluid Boundaries and Limited Authorities**: Due to the nature of technologies within a smart city, the political and organisational boundaries are ever-changing over time. Having someone responsible for this risk management can be challenging to achieve in itself.

4.4.1 Training and Awareness

When developing smart cities there is an essential requirement for education and development of training policies. Whereas FIPPs, privacy-by-design and security-by-design deliver practical results, they would not be utilised to their full potential without supporting users' understanding of the technology. Users should also be provided with guidance on how they are impacted by systems and how to use them, whilst demonstrating protection for their privacy and security. To support this area, four types of national education and training programmes are highly encouraged:

- General Education Programme: this target audience is the general public. It provides an overall understanding of how to protect against privacy and security threats and the steps required to take to ensure this is put physically into practice. It also covers the technologies embedded within a smart city and the privacy and security risks threats that are linked
- School Children Education Programme: This focuses on educating school children on data privacy and how data is generated on them. This can include avenues such as social media usage and understanding the risks of what personal data is made available.
- Local Authority Education Programme: This focuses on the education for local authority staff and what their responsibility is when it comes to data protection and how to assess and mitigate privacy and security risks
- Technology Companies Education Programme: This mainly focuses on SMEs and new businesses that may not have the education programmes that a large established company would typically have. This provides opportunities for these businesses to understand what best practices to implement and what their responsibilities are to the public and customers.

4.4.2 The Governance Framework and Management Support

Governance provides a framework where decisions can be made, and a strategic direction is set and agreed upon. Governance also includes the practical implementation and monitoring of regulations. Management provides the driving force to achieve the goals in implementing and running the necessary services. Both these aspects complement and support the set-up and maintenance of technologies and systems while adhering to legal and regulatory requirements.

Today it is predominantly found that smart cities have been developed with minimal support in this area, resulting in clear gaps in privacy and security management. To gain the trust of the users and facilitate faster adoption of smart city technologies this approach must be replaced with a more organised one to maintain direction and achieve strategic objectives. Oversight and compliance require incorporation into existing (and future) practical deployments, managing governance and day-to-day delivery with an emergency response team to hand to provide that trust for citizens. Putting these factors in place will hugely improve a smart city's potential, promote key learning to support future sustainable development.

4.4.3 Smart City Cyber Incident Response Teams (SC CIRTs)

Smart City Cyber Incident Response Teams (SC CIRTs) play a critical role in incident management, mainly managing cybersecurity incidents such as hacking, data theft, system disruption or termination. They are a team of key individuals that manage incident response and security. Due to the nature of their job, they have a range of planned scenarios with planned responses to effectively deal with and manage an incident to best practice. The scenarios are revised regularly to ensure they remain current and effective and also highlight any new risks that need to be planned for.

4.4.4 Comply and Manage

Some key principles support comprehensive cooperation and coordination in governance within a smart city. These are:

- **Accountability**: Due to the complex nature of smart cities, it must be clear where accountability lies, for example in the storage of data, incident management, education and risk assessments.
- **Collaboration**: To successfully develop and maintain a smart city it is key for experts in different areas to collaborate or utilise collaborative models [33] powered by technology to identify and manage risks and create mitigating controls. Residents within the smart city should also have involvement as its important concerns are addressed, and everyone's needs and best interests are understood.
- **Leadership responsibility**: It is a requirement to have a clear leadership structure that has responsibility for the city's security and resiliency. The elected officials are to understand the needs of their residents while also being aware of privacy and security risks.
- **Trust and Transparency**: To create trust within a city between residents, businesses and local authorities, transparency is key. Not only do residents have a right to be able to see relevant data but it also raises awareness on cybersecurity issues and best practices going forward.

From these principles, the following recommendations can be actioned led by the advisory board:

1. Coordinate Forums and Promote Collaboration: By installing regular forums to raise concerns, share best practices and share information promotes collaboration with a range of expertise and across different perspectives to achieve the best outcomes.

2. Organise Roadshows: Leadership needs to organise roadshows or roundtable events to raise topics and debate and resolve issues around the smart city. Early engagement within communities and throughout will ensure that smart technologies do not worsen existing discriminations and inequalities.

3. Provide Transparent and Frequent Communication: By communicating key messages, recent changes, decision making, and concerns will build trust within the community through its transparency.

5 On the Development and Creation of a Secure and Resilient Smart City

During the early stages of smart city planning and development, it must be a priority to implement security and resilience. A roadmap should be developed by smart city managers in these early stages. Regarding the CyberSmart CRPG Framework, "Connect", "Resist", "Protect" and "Comply and Manage" should be aspects of consideration when building the roadmap.

It is a common error that security, privacy and resilience are not taken into consideration until it is too late. It is also a problem if such a matter is not strategically planned for by the city's senior management board. Figure 10 illustrates an example for such a roadmap with three clear functions;

- Provides a starting point: this ensures individuals understand what the current state of play is and supports to process of achieving an end goal.
- Engagement Awareness: This supports the collaboration of key stakeholders' input by creating an avenue to share ideas, raise key issues and identify solutions.
- Environmental Factors: This is where internal and external factors come into consideration, such as changes to laws and regulations, technology developments and changes to consumer needs.

Each city is unique, consisting of different sizes, cultures, style of governance, priorities and needs. However, even though there are varying elements within a smart city, a roadmap should still follow a set of key elements [18]:

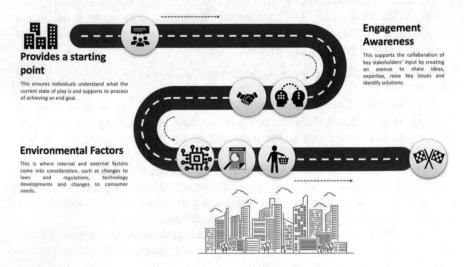

Provides a starting point

This ensures individuals understand what the current state of play is and supports to process of achieving an end goal.

Engagement Awareness

This supports the collaboration of key stakeholders' input by creating an avenue to share ideas, expertise, raise key issues and identify solutions.

Environmental Factors

This is where internal and external factors come into consideration, such as changes to laws and regulations, technology developments and changes to consumer needs.

Fig. 10 Roadmap of a secure and resilient smart city

1. **Create Smart City Objectives**: This element focuses on what the overall vision is for the city, what are the reasons for becoming a smart city and what are the key impacts this will have.
2. **Create a Strong Governance Structure**: Ensure security and resilience issues are handled with a collaborative approach to achieve the best possible outcomes. Governance should provide transparency to members on key issues and actions.
3. **Complete Security and Resiliency Reviews**: Create adequate security, resilience and data protection policies. Understand what gaps exist within the city and develop controls to close these gaps. Provide education to the public on security awareness.
4. **Effective Risk Assessment Completion**: By completing risk assessments on each of the smart city's domains allows for adequate contingencies to be implemented should the main technologies or devices become unavailable. During the risk assessment process, a full impact assessment should be completed to understand the impacts, whether these are financial, reputational or policy-related and how best to prevent or minimise them.
5. **Identify Critical Areas and Develop Supporting Processes**: Complete a review of the cities critical areas and understand what processes or third parties support this. This ensures high-quality data which is especially important for example, in the event of system failure the impacts and required stakeholders are ready to hand.
6. **Stakeholder Collaboration**: Ensure a range where key stakeholders have identified that offer different areas of expertise to build the best smart city possible.
7. **Technology Investments**: To build a robust smart city, money must be invested into technologies and devices. Detailed risk assessments should be completed on the implementation of different systems to understand the benefits and risks to make an informed decision.

6 Conclusion

Smart cities have real value to offer, however, there are also serious risks that can impact privacy and therefore the safety of its citizens. Taking a proactive approach to developing and maintaining a smart city will ensure that the full benefits are appropriately utilised, and adequate risk management is built in. In this study, several challenges have been defined and discussed such as operational risk management within smart cities, management structure and key stakeholders complexity, and lack of confidence and trust in new devices and technology. Therefore, a CRPG framework have been proposed and discussed to analyse and map risk to mitigation activities with a particular focus on data protection and data privacy.

Furthermore, as cities vary, needs and wants can vary enormously, therefore, when developing a smart city, collaboration should be facilitated through appropriate frameworks to acknowledge the size, sectors, cultures, history, and economic strength across a diverse steering group. Having clear objectives in place on what improvements are a priority, provides focus on the main reasons to build the smart city. For example, if there is a growing concern for the environment, clear objectives around this need to be established, and strong monitoring and performance tracking put in place. Likewise, educational needs must be planned for the public to raise awareness and ensure participation. The more buy-in that is received from the public, the faster benefits can be realised, and the next steps put in place to benefit all across the longer-term vision.

Following the outcomes of this study, we also wanted to shed light on the emerging concept of Digital Twins [5], a cutting-edge technology that we expect will act as a guide in effectively planning for the future. These technological advances support a city to provide enhanced services, recourses and improve assets. This promotes a better quality of life and provides opportunities for growth and development securely. By embedding digital twins within a smart city framework, we anticipate improved responses to disasters and emergencies.

References

1. Coletta C, Kitchin R, (2017) Algorhythmic governance: regulating the 'heartbeat' of a city using the Internet of Things. Big Data Soc 4(2). https://doi.org/10.1177/2053951717742418
2. Ahmadi-Assalemi G, Al-Khateeb HM, Maple C, Epiphaniou G, Hammoudeh M, Jahankhani H, Pillai P (2021) Optimising driver profiling through behaviour modelling of in-car sensor and global positioning system data. Comput Electr Eng 91:107047. https://doi.org/10.1016/j.compeleceng.2021.107047
3. Ahmadi-Assalemi G, Al-Khateeb H, Epiphaniou G, Maple C (2020) Cyber resilience and incident response in smart cities: a systematic literature review. Smart Cities 3(3):894–927. https://doi.org/10.3390/smartcities3030046
4. Albishi S, Soh B, Ullah A, Algarni F (2017) Challenges and solutions for applications and technologies in the internet of things. Procedia Comput Sci 124:608–614. https://doi.org/10.1016/j.procs.2017.12.196
5. Ahmadi-Assalemi V, Al-Khateeb H, Maple C, Epiphaniou G, Alhaboby ZA, Alkaabi S, Alhaboby V (2020) Digital twins for precision healthcare. In: Jahankhani H, Kendzierskyj S, Chelvachandran N, Ibarra J (eds) Cyber defence in the age of AI, smart societies and augmented humanity. Springer International Publishing, pp 133–158. https://doi.org/10.1007/978-3-030-35746-7_8
6. Lisdorf A (2019) Demystifying smart cities: practical perspectives on how cities can leverage the potential of new technologies. Springer. https://doi.org/10.1007/978-1-4842-5377-9
7. Chang V, Sharma S, Li C-S (2020) Smart cities in the 21st century. Technol Forecast Soc Change 153:119447. https://doi.org/10.1016/j.techfore.2018.09.002
8. Kumar TV (2020) Smart environment for smart cities. Springer, pp 1–53. https://doi.org/10.1007/978-981-13-6822-6
9. Caragliu A, Bo CD, Nijkamp P (2011) Smart cities in Europe. J Urban Technol 18(2):65–82. https://doi.org/10.1080/10630732.2011.601117
10. Sikora-Fernandez D, Stawasz D (2016) The concept of smart city in the theory and practice of urban development management. Romanian J Reg Sci 10(1):86–99
11. Al-Turjman F (2019) The road towards plant phenotyping via WSNs: an overview. Comput Electron Agric 161:4–13. https://doi.org/10.1016/j.compag.2018.09.018

12. Al-Turjman F, Lemayian JP (2020) Intelligence, security, and vehicular sensor networks in internet of things (IoT)-enabled smart-cities: an overview. Comput Electr Eng 87:106776. https://doi.org/10.1016/j.compeleceng.2020.106776
13. Dincer I, Acar C (2017) Smart energy systems for a sustainable future. Appl Energy 194:225–235. https://doi.org/10.1016/j.apenergy.2016.12.058
14. Robles T, Alcarria R, Martín D, Morales A, Navarro M, Calero R, Iglesias S, López M (2014) An internet of things-based model for smart water management, pp 821–826. https://doi.org/10.1109/WAINA.2014.129
15. Snellen D, Hollander GD (2017) ICT's change transport and mobility: mind the policy gap! Trans Res Procedia 26:3–12. https://doi.org/10.1016/j.trpro.2017.07.003
16. Jahankhani H, Kendzierskyj S, Jamal A, Epiphaniou G, Al-Khateeb H (2019) Blockchain and clinical trial: securing patient data. Springer. https://doi.org/10.1007/978-3-030-11289-9
17. Dehghantanha A, Choo KKR (2019) Handbook of big data and IoT security. Springer. https://doi.org/10.1007/978-3-030-10543-3
18. EastWest Institute (2018) Smart and safe—risk reduction in tomorrow's cities. https://www.eastwest.ngo/sites/default/files/ideas-files/ewi-smart-and-safe-cities.pdf
19. Ismagilova E, Hughes L, Rana NP, Dwivedi YK (2020) Security, privacy and risks within smart cities: literature review and development of a smart city interaction framework. Inform Syst Front. https://doi.org/10.1007/s10796-020-10044-1
20. Kitchin R, Dodge M (2019) The (In) security of smart cities: vulnerabilities, risks, mitigation, and prevention. J Urban Technol 26(2):47–65. https://doi.org/10.1080/10630732.2017.1408002
21. Ahmadi-Assalemi G, Al-Khateeb HM, Epiphaniou G, Cosson J, Jahankhani H, Pillai P (2019) Federated blockchain-based tracking and liability attribution framework for employees and cyber-physical objects in a smart workplace. pp 1–9. https://doi.org/10.1109/ICGS3.2019.8688297
22. Cui L, Xie G, Qu Y, Gao L, Yang Y (2018) Security and privacy in smart cities: challenges and opportunities. IEEE Access 6:46134–46145. https://doi.org/10.1109/ACCESS.2018.2853985
23. Andrade RO, Yoo SG, Tello-Oquendo L, Ortiz-Garcés I (2021) Chapter 12—Cybersecurity, sustainability, and resilience capabilities of a smart city. In: Visvizi A, Hoyo RPD (eds) Smart cities and the un SDGs. Elsevier, pp 181–193. https://doi.org/10.1016/B978-0-323-85151-0.00012-9
24. Ersotelos N, Bottarelli M, Al-Khateeb H, Epiphaniou G, Alhaboby Z, Pillai P, Aggoun A (2021) Blockchain and IoMT against physical abuse: bullying in schools as a case study. J Sens Actuator Netw 10(1):1. https://doi.org/10.3390/jsan10010001
25. Torra V (2017) Data privacy: foundations, new developments and the big data challenge. Springer
26. International Commission of Jurists (1967) The right to privacy: working paper. Stockholm, Sweden, May 22–23. https://www.icj.org/nordic-conference-of-jurists-the-right-to-privacy-working-paper-stockholm-sweden-may-22-23-1967/
27. United Nations (2015) Universal declaration of human rights. https://www.un.org/en/universal-declaration-human-rights/
28. Haughey H, Epiphaniou G, Al-Khateeb HM (2016) Anonymity networks and the fragile cyber ecosystem. Netw Secur 2016(3):10–18. https://doi.org/10.1016/S1353-4858(16)30028-9
29. Solove DJ (2005) A taxonomy of privacy. Univ Pa Law Rev 154(3):477–564.
30. Irshad M, Al-Khateeb HM, Mansour A, Ashawa M, Hamisu M (2018) Effective methods to detect metamorphic malware: a systematic review. Int J Electron Secur Digit Forensics 10(2):138–154. https://doi.org/10.1504/ijesdf.2018.090948
31. Woensdregt JW, Al-Khateeb HM, Epiphaniou G, Jahankhani H (2019) AdPExT: designing a tool to assess information gleaned from browsers by online advertising platforms, pp 204–212. https://doi.org/10.1109/ICGS3.2019.8688328
32. Haber E, Tamò-Larrieux A (2020) Privacy and security by design: comparing the EU and Israeli approaches to embedding privacy and security. Comput Law Secur Review 37:105409. https://doi.org/10.1016/j.clsr.2020.105409

33. Al-Husaini Y, Al-Khateeb H, Warren M, Pan L, Epiphaniou G (2020) Collaborative digital forensic investigations model for law enforcement: oman as a case study. Security and organization within IoT and Smart Cities. CRC Press, pp 157–180. https://doi.org/10.1201/978100 3018636-9

A Descriptive Analytics of the Occurrence and Predictive Analytics of Cyber Attacks During the Pandemic

Emmanuel Folusho Adeniran and Hamid Jahankhani

Abstract The SARS-CoV-2 (Severe acute respiratory syndrome coronavirus 2) universally and commonly known as COVIT-19 and the across-the-board lockdown measures are having complex and unforeseen effects on intricate social domains, which includes opportunities for offline and online crimes. For some months since March 2020, most countries in the world if not all has been in one lockdown or the other. The lockdown measures have brought increased fear, anxiety, and depression. As if that is not enough, it has also forced the entire world populace to embrace the emergency use of technology to carry out their job functions which was unplanned for in a sit-at-home and work-from-home situation. All these have rendered vast majority of individuals and general populace vulnerable to cyber enabled and cyber dependent crimes. Anytime there is spike in number of crimes committed within a specific period, analysists and researchers seek to know the factor responsible for the increase. In this research, attempt is made to establish the relationship between the COVIT-19 pandemic and the increase in cybercrime. To achieve this, Regression Analysis is being used to study the relationship between the target class, attack class and the cybercrime occurrence. Furthermore, projection is also made using Time Series to see how the situation will look like if this lockdown that some have accepted to be a new-normal, continues like this (God forbids that).

Keywords Time series · Cyber-attack · SARS-CoV-2 · COVIT-19 · Cyberwarfare · Routine activity theory · Cyber criminology

1 Introduction

The pandemic brought about by the Covit-19 has been an unprecedented and remarkable event. The pandemic has altered the lives of billions of people all over the world bringing about what is commonly known as a New-Normal when it comes to mode of operation, way of life, work, and societal norms. Economic activities were almost

E. F. Adeniran · H. Jahankhani (✉)
Northumbria University, London, UK
e-mail: Hamid.jahankhani@northumbria.ac.uk

brought to a total halt with oil prices per barrel dropping to as low as bellow $0 per barrel in April 2020 whereby Oil producers had to pay money to disburse crude oil [1].

Aside from the impact on all spheres of life that it had, it has also opened door for the influx and occurrence of cybercrime. The pandemic has brought about a set of increased and distinct crime via web which has negatively impacted on all businesses and society at large. It exposes the various vulnerabilities of individuals, governments, academic institutions, and various organisations during the period. No thanks to the sit-at-home and work-from-home (work remotely) rules and conditions popularly known as social distancing enforced on the general citizenry all over the world.

The fear, worry, anxiety and ignorance or lack of knowledge of vast majority of people all over the world has been majorly capitalised on and exploited by criminals to perpetrate this erroneous crime [2]. Since the beginning of this pandemic, there has been an upsurge of cybercrimes in different dimensions, at different levels and using various methods.

This research attempts to carry out a quantitative analysis of the occurrence and go further to do a near-perfect prediction that will help government establishments, key players, stakeholders, and the world at large in web security to plan and strategize on estimated number of cybercrimes in the nearest future and adequate plans to counter it. This paperwork seeks to bridge this gap.

The web has become a cyberwarfare where battle goes on between the good and the bad constantly. Yet, in a world that is fast becoming a global village, the web is unavoidable. Further still, the warfare has now been aggravated by the Covit pandemic. What is a Pandemic? According to the World Health Organisation, pandemic can be defined as an epidemic that has spread all over the world or over a very wide area coverage cutting across national and international boundaries and generally affecting large percentage of the inhabitants [3].

One of the tragedies of this lockdown is that it has come like a storm and world economies and stakeholders in the security sector has been caught unaware and unprepared.

In a rare time like this described as the New-Normal, in my own words, cyber-crimes carried out in times like this can be described as Opportunistic crimes. This is because crimes like this via cyber route is done by taking advantage of vulnerable and unsuspecting members of the public. As we all know, the world environment is tense, everyone going about with fear and vigilance. No thanks to the social distance rules and conditions being enforced upon the people all over the world in this period. This vulnerability comes as a window of opportunity to the ever-ready men of the underworld whose profession is this nefarious act called cyber-crime [4].

Kathy Macdonald in the book she authored made a vital point that the world at large should put into serious consideration [2]. She is of the opinion that cybercrime cannot be curbed without the following.

- Creation of a well detailed awareness
- Effective prevention methods

- Appropriate response strategies.

For a responsible and up-to-the-task government and stakeholders there should be a proper system of analysing and presenting this crime in such a way as to reveal the enormity and severity of this crime. There should also be a framework that projects into the future of cybercrimes in the form of predictions and forecasting. Michael Levi ascertained that there exist a quandary or challenge of what system of measurement or metrics are suitable for evaluating the threat and damage from cybercrimes [5].

This piece of work or project aims at providing a framework that will meet the demands of this agitation. This will be carried out by using a non-conventional approach.

This project seeks to follow or adopt a different approach from the common method of analysis amongst other research works done in this field of cybercrime analysis.

The conventional method has been Routine Activity Theory. But this piece of work seeks to follow a different approach by going the way of a quantitative analysis. Two major statistical tools will be used for the predictive analysis, namely-Time series, and Regression analytics.

The major significance of this project is to come up with a framework that will give thorough and precise insight into the immediate future of cybercrime in a time like this.

2 Literature Review

2.1 Introduction

The heightened and increased rate of occurrences of cybercrime worldwide has got to a level that demands serious and urgent escalation. Activities of the men of the underworld perpetrating this nefarious act has gotten to a perilous magnitude resulting into loss of hard-earned funds, properties, and valuables. Worst still, is that it is beginning to degenerate to doubt and loss of trust in the use of technology. Yet researchers and stakeholders have been kept on their toes in fighting this type of crime.

With the way cybercrime rate has sky-rocketed during this covit-19; threatening members of the public, individuals, families, businesses and even the state, it has become imperative and necessary for players and stakeholders in the security sector to have various framework that will help report this menace and create more awareness about it [6].

In this chapter, a critical look is taken at few out of the various works done by researchers (both academic and non-academic) and stakeholders to curb this crime in terms of profiling.

This will be considered under the following sub-topics.

- Definitions
- Theories and framework
- Geographical Influences
- Comparison of Cybercrime with Traditional crime
- Cybercrime profiling Using Routine Activity Theory (RAT)
- Cybercrime profiling with Time series
- Cybercrime profiling with Regression analysis.

2.2 Definition

Before going deeper into this very important and crucial topic in a rare time like this known as the new-normal, it will be a good way to start by considering the definition of cybercrime.

The pair of Eric Rutger Leukfeldt and Majid Yar defined cybercrime as offences and crimes committed with the involvement and dependence on the utilisation of modern-day communication technologies [7].

Crown Prosecution Service (CPS), an independent body saddled with the responsibility of prosecution of criminal cases that has been investigated by the police and other investigative bodies for England and Wales described cybercrime in their cybercrime prosecution guidelines as an umbrella expression employed to illustrate two very related, but yet discrete varieties of crime [8]. They are:

- **Cyber-dependent crimes**—crimes carried out using Information and Communications Technology (ICT) devices. In this scenario, the devices playing the role of both the tool for perpetrating the crime, and the victim/target of the crime. Example includes when a malware is being developed and disseminated for financial gain, when hacking is being carried out to steal, malicious script being deployed to gain illegal access to a network or system, viruses being deployed to a system or network to damage, distort and alter data.
- **Cyber-enabled crimes**—Conventional crimes whose impact/damage can be maximised using computers, network of computers or various techniques in ICT (such as stealing of data and frauds that are cyber-enabled).

Iesh Chandraa and Melissa J. Snowe defined cybercrime as an act that engages computer technology to carry out a crime. They went further to talk about what they described as four pillars of the taxonomy of cybercrime as:

- Mutual exclusivity
- Structure
- Exhaustiveness
- Well-defined categories.

All these put together gives a theoretical basis for the taxonomy of cybercrime [9].

2.3   Theory and Model

When it comes to the study of criminology, different theories and models have been postulated and used to profile cybercrime. A team of researchers carried out a glossary look at various theories that generated different types of models for cybercrime profiling [10]. According to this team, due to the fact that there are various reasons, attributes, components, targets, methods and dimensions of Cybercrime, It has necessitated various approaches to cybercrime profiling. Various theorists have come up with various framework to get to the root of cybercrimes worldwide. They considered some of the theories and their corresponding mappings to cyber-enabled and cyber-dependent crimes (See Table A).

Table A   Theories on criminology and its corresponding mapping [10]

Theory	Criminology concept	Mapped to cypercrime Perspective	Ottender
Social learning theory	Criminal behavior is acquired through observational learning	Cybercriminal conduct crime by imitation and modelling from others (etc: Hacking)	Offender
Cultural lag	Failure to develop social consensus on appropriate application of modem technology lead to a breakdown in social settings	Any application/technology designed with no regard of protection lead to the human being the weakest link	Guardianship
Digital drift theory	Internet interactions, which requires no face-to-face and borderless could drift non-criminal towards criminality	Some offenders get drifted into becoming cyber criminals due to internet pseudo-reciprocal environment (etc: Child pornography)	Offenders
Space transition theory	Explanation about the nature of the behavior of the persons who bring out their conforming and nonconforming behavior in the physical space and cyberspace	Factors such as offenders behavior; the social settings they live in and the internet features lead to cybercrime	Offender
Situational crime prevention	Crime can be reduced by altering situations rather than an offender disposition	Implementation of the defender/guardianship system, reduce likelihood of crime	Guardianship

Crane Hassold invented a very direct and very comprehensive formulae for individuals and stakeholders, when it comes to fraud in cybernetics and computer technological fields to identify a malicious mail. In his email security blog publication titled "Threat Taxonomy- A working framework to describe cyber-attacks" The threat taxonomy illustrates a threat that is communication-based looking at it from different dimensions and perception [6]. He took it from the point of view that for cybercrime to be properly analysed, some important questions need to be raised. He commenced by looking at the classification of the message in question. Is the message a malicious one or not? Can it be referred to as a grey mail or it could be grouped under spam. In his formulae, he went ahead to consider the Genuity or authentication of the source or sender. His theory attempts to establish that, the best way to fight crime by asking the right set of questions.

Gregory Epiphaniou and four other researchers did a group work and analysed covit-19 from the cybersecurity point of view. Their analysis considered the pandemic from the perspective of cyber-crime and underscores the range of cybercrime and frauds recorded worldwide [11].

In their research, a critical study was made based on the context or within the premise of a symbolic world event to reveal the mode of operation and execution of cybercrimes. This piece of work reveals how cybercrimes became much more rampant and ubiquitous sequel to the period between when the covit-19 pandemic first broke out in China and the first cyber-attack that has to do with Covid-19. Ever since then, it has steadily increased to an extent that some days will not pass without about 3–4 cases of unique and distinct cyber-attacks were being recorded. The study went ahead using the UK as a specimen to illustrate how cyber-fraudsters take advantage on major events and state activities and announcements to craftily orchestrate or perpetrate cybercrimes.

Apart from the geographical influence that can impact cybercrime, Asmir and her group went ahead to build on the algorithm of Criminal Geographic Targeting (CGT). This algorithm illustrates the mathematical correlation or relationship that exist between defaulter's travel and the likelihood of committing the offence to come up with the most likely or possible area of the criminal's hideout or abode using the creation of a surface of jeopardy or geographical profile [12]. They based their assumption on the fact that there are numerous mathematical techniques that trigger the current strategies of location-based profiling categorized into two broad classifications namely Spatial Distribution Strategies (SDS) and Probability Distance Strategies (PDS).

One of the most common and conventional theory to explain, analyse and give a detailed description of this menace called cybercrime is Routine Activity Theory (RAT). Using this theory, Jungkook An and Hee-woong Kim revealed that for a cybercrime to be successful, there are three major factors that must be in place [13].

- A prospective criminal
- A vulnerable fellow or system
- Unavailability of a system to checkmate cybercrime.

In a proper context of cybercrime, prospective criminals are the major players and stakeholders in the crimeware market. They are the producers of the various crimeware and those who patronise them. Vulnerable persons or systems refers to the unsuspecting persons who are ignorant of the devices and activities of the men of the underworld. In this covit 19 period, in all part of the world, people are made to work from home. Various board meetings all over the world had to be done online using platforms like Zoom, free conference call, TeamViewer, and others. In fact, during this covit-19 pandemic, report has it that daily usage of zoom rose from 10 million to over 200 million [14]. The third major factor that must be in place for a cybercrime to be successful is unavailability of a system that is well-defined and effective for creating awareness and curbing cybercrime [13].

Lohrmann Dan- an author, a keynote speaker, the current Chief Security Officer (CSO) and Chief Strategist-Security Mentor Incorporated, while researching and analysing the Ransomware During Covid-19 gave some seven signals or red flag to watch out for during this period [4]. In his analysis and considering the increase in covid-19 cases and corresponding cyberattack, he considered the attack data sourced from VMware carbon black cloud to determine the shift to remote work. In the fourth paragraph of his report, stated how cyber criminals and attackers have advanced in their operations, timing of their operations and their target-industries and victims [4]. The seven red flag he gave are as follows:

- Active directory revealing multiple login failures
- Phishing emails coming with unknown and strange domains
- Some string of questions being raised about a particular machine
- Emergence and utilization of security tools in environments where they were not assigned
- Various networks being hit by Brute force attacks
- Sudden Redirection of traffic to questionable points or systems on the dark web.

Although in his final thoughts, he opined that the focus of cyber attackers and hackers using malware has been on hospitals and research institutions more than governmental institutions and other sectors [4], this is not entirely true as there has been more successful ones in some other areas. A good mathematical framework or model will be able to indicate/reveal and support/fault this claim [15]. Though with a holistic ground, this is one of the areas this project works seeks to address.

Another work worthy of note was done by Erin Mikolai in his project he titled "Cybercriminals Exploitation of the Corona Virus Pandemic". In his piece of work, he analysed this nefarious crime during this rare time by looking at the various effects and impact of the pandemic that then catalysed the increase of cybercrime [16]. In his research, he was of the opinion that the pandemic is not a direct cause of increase in cybercrime. But rather, the pandemic and the responses to it influenced lots of various aspect of our life and society which then resulted in the creation of vulnerabilities in the life of the various victims.

2.4 Geographical Influence

A group of researchers comprising of about four of them did a research on Geographic profiling for serial cybercrime investigation [12]. In their research work, they remarked that modern day cybercrimes are much more complicated and challenging to successfully investigate and prosecute when compared with traditional crimes. They believe that there is a strong relationship and link between the geographical location and the cybercrimes committed.

Their focus was beamed towards the practicability of applying the technique of geographic profiling to cybercrimes investigation. Their research is primarily based on the premise or assumption that for almost all types of crimes committed via web, steps taken to carry out these nefarious crimes are not random [12]. For instance, the target class, location of crime, method of attack, information gathering, and reconnaissance all follow a pre-determined logic or reasoning which could bring about a breakthrough in the crime detection and curbing. Testing the effectiveness and functionality of a geographical profiling has been done on actual samples of crimes committed via web gathered by security agencies and practitioners. Their research was aimed at applying the theory of geographical profiling to investigate crimes committed via internet which involves the natural environment. They focused on just types of crimes namely—Spear phishing and skimming of credit card. They used a tailor-made Geo-Crime geographic profiling application/software created to help in the plotting, mapping, four-dimensional and statistical analytics of cyber enabled and cyber-dependent crimes patterns. They claimed that, though under some specific conditions, the result of their study have established the feasibility of adopting geographic profiling to investigate cybercrimes. Their selling point is the fact that geographic profiling is most effective and result-yielding in cases where the hacker conceals his identities and activities with the use of internet [12]. In their conclusion, they argued that in present day web environment, it is necessary to new and dependable methods of investigating cybercrimes and cyber offenders profiling. They believe it is time to start considering cybercrimes as similar or synonymous to traditional crimes (the likes of knife crime, fraud, terrorist attacks and bombings, rape, and other illegal actions) where series of crime prevention techniques have been employed over a long period of time. In their own view, geographical profiling should be one of the leading techniques in that class. Their research also reveals that some group of criminally minded offences conform to the requirements for the application of geographic profiling which is based on the premises that those who commit this nefarious group of crimes via web follows a certain pattern of logic [12, 17].

Research on cybercrime profiling spreads his tentacles to Nigeria- a country in the sub-Saharan region of Africa. In his research works, Suleman Ibrahim considered Social and contextual taxonomy of cybercrime as a Socioeconomic theory of Nigerian cybercriminals [17]. The aim of his project works was to establish the idiosyncrasies and peculiarities of cybercrime in Nigeria and whether these indicate challenges with prevailing nomenclatures and classification of cybercrime. It anchors upon a fundamental theory or assumption of categorisation coupled with the theory of motivation, to offer a three-way conceptual structure for categorizing cybercrime relationship. This paper works also claims that cybercrimes are inspired and motivated by three likely premise or factors. They include:

- Geopolitical factor
- Psychosocial factor

- Socioeconomic factor.

Although this piece of work questions the source of statistics used to bring about the predominance of cybercrime culprits across the globe, it gives room for innovative means of the utilisation of the capacious modifications of this nefarious crime referred to as cybercrime.

In Tandem, it provides a brighter formation of the idea and understanding of cybercrime in the sub-Saharan region of Africa and elsewhere using Nigeria as a case study.

This is so because the geopolitical, psychosocial, and socioeconomic factor also applies online as they are known to influence offline [17]. Ben Stickle and Marcus Felson seem to share the idea that crimes are influenced by these three key factors have great impact on the occurrence of crime during this pandemic [18]. They also share the opinion of Ibrahim Suleman that crime rate varies from one area to another even during the lockdown introduced by this pandemic [17]. They (Ben and Marcus) observed that researchers in the field of criminology have the propensity to draw inference from a few cases to describe most crimes while undervaluing the massive and influential specificity in culprit's decision making. According to their research works, specificity cannot be over-emphasized when profiling crime during pandemic as it gives room for the clarity of understanding of the changes that are nuanced in nature. This according to them is one of the reasons why Cybercrime is at alarming rate during this pandemic. In their opinion, crime specificity will unravel the opportunity structure, unique skills, and mode of operation of the offender. This in turn will give room for the better understanding of the patterns of the crime and various techniques adopted [18]. For instance, the variation in daily routine activities in the rouse of the lockdown introduced by the pandemic have a propensity to reduce the population in areas of the metropolis considered to be non-residential, though boosting the population in regions or areas that are residential [18].

After giving a critical look to the various analysis of cybercrime during this pandemic, it will be discovered that almost everyone followed the bandwagon of same methodology which is Routine Activity Theory (RAT) except for some very few. Jordanne Morrow in his findings concluded that before a theory can be considered usable, it must carry the elements of empirical validity to an appreciable extent [19]. The Routine Activity theory which was developed by Cohen and Felson in 1979 postulates that before a crime can be actualised, three key factors must be in place. Firstly, there must be potential offender, a suitable target and lack of an effective guardian [20].

According to Glenn Kitteringham, the attention or concentration of Researchers, stakeholders, and players in the environmental crime is towards location of prospective fraudulent or criminal activity and endeavour to analyse crime based on environmental influences [20].

2.5   Comparism with Traditional Crime

While, this COVIT-19 pandemic and lockdown has brought about great increase in the number of cybercrimes, research has also revealed that there has been a great reduction in the occurrence of traditional crime [18].

According to them, With the current decline in traditional crime at such a substantial rate and numerous of the commonly attributed scenarios influencing crime remaining consistent and in some scenarios are escalating or diminishing in opposite direction of what is regarded as the consensus of factors that enable crime, most of the criminological theories are seen to be failing or not entirely true.

To illustrate this sudden drastic change, it is believed that the extent and type of crime variations or changes experienced during the lockdown will serve as a reliable test or proof for the numerous theories that endeavour to describe the study of criminal behaviour. Eventually, this basically and obviously transpiring experiment will increase our knowledge of crime and behavioural science as no other occurrence has succeeded in achieving this at a time that data on criminology were readily obtainable [18].

Hilton Collins, a writer, and an ex-staff of Government Technology is of the opinion that a distinctive and different approach to analysing crime may enable the police personnel to forecast or predict crime. In his article, Hilton reported the approach taken by Professor Joel Caplan, an assistant Professor in the School of Criminal Justice, Rutgers University. Caplan teamed up with two of his colleagues from the University- professors Leslie Kennedy and Eric Piza to research and explore the risk terrain modelling [21].

2.6   Cybercrime Profiling with Time Series

Hilton in this methodology, uses a region and merges its record of crime with data on the grounds of physical characteristics and local behaviour to build a trend map of locations with the highest crime threat or risk. According to Caplan, in his words "It paints a picture of those underlying features of the environment that are attractive for certain types of illegal behaviour, and in doing so, we're able to assign probabilities of crime occurring," [21].

A group of researchers made of five members from the department of criminology, University of Manchester made effort to explore on this very important research area [22]. They did a time series analysis of cyber related and cyber-enabled crime of May 2020 and that of May 2019. They went further to compare the results from both analyses to establish that cyber dependent and cyber enabled crime has been greatly increased by the COVIT-19 pandemic and lockdown using Time series analysis. In their piece of work, they pointed out some very important factors which can or has greatly affected the accuracy and reliability of the data used for this research work. The following points were deduced from their observation:

- Possibility of lockdown measures impeding or hindering the accurate and complete reporting of cyber enabled and cyber-dependent crimes during the pandemic
- Possibility of measuring errors from the data source
- The responses are more of opinion responses (open ended questions) which is qualitative and for it to be relevant for this study it must be coded into close ended questions.
- Missing values (responses) on the authors (those bodies through which the crimes were perpetrated) involved in the criminal acts.
- Tracking the criminal acts as it was carried out in different counties.
- Possibility of data poisoning.
- The integrity of the reporting body cannot be said to be tantamount to the data integrity. As a result, the integrity of the data cannot be verified to a certain level or proved to be one hundred percent accurate.

This group went further to carry out the analysis using time series analysis to compare the periods under consideration to drive home their claims.

Though as impressive as their work seems, the scope of their research was limited to just the comparison between this period.

2.7   Cybercrime Profiling with Routine Activity Theory (Rat)

Erik Rutger Leukfeldt and Majid Yar did a research work titled "Applying Routine Activity Theory to Cybercrime: A theoretical and Empirical Analysis".

In the piece of work, the main question raised was if Routine Activity Theory can be employed as an analytical model, framework, or basis to research cybercrimes. In their words, they eventually concluded that "both theoretical analysis and the analysis of empirical studies have thus far failed to provide a clear answer." [7]. To avoid the various shortcomings of most Routine-Activity-Theory based studies, the multivariate analysis depicted in this study endeavour to prevent most of the limitations of other works that are RAT-dependent. A large population of about 9,161 was adopted for the research which was tested on six different cybercrimes and the following was studied:

- Visibility
- Accessibility
- Effect of Value
- Guardianship on victimization.

In their result, it was deduced that some elements of RAT are more relevant than some other ones. It also reveals that in cybercrime victimization.

Visibility undoubtedly performs a crucial role. When it comes to capable personal guardianship and accessibility, they reveal contrasting results. Technical capability in guardianship and value represents little or no significance or impact on victimization of cyber-enabled and cyber-dependent crimes.

2.8 Cybercrime Profiling with Regression Analysis

Olayinka and co. examines the relationship between risk governance and cyber-crime using hierarchical regression approach [23]. They went ahead to investigate the impact and effect risk governance have on the cybercrime occurrence using some selected firms in the financial institution of a sub-Saharan country of Africa- Nigeria to be precise.

To accomplish this, they took a randomly selected sample of 50 enlisted companies in the financial sector for a period of five years (2013–2017) and they employed the use of hierarchical regression analysis to test the measure of relationship and influence of some factors (independent variables) on cybercrime. The factors used as independent variables includes.

1. Risk Governance
• Chief Risk Officer_centrality
• Size of Board Risk Committee
• Enterprise Risk Manager_index chief
• Board Risk Committee_ activism
• Risk Officer_presence
• Board Risk Committee_independence
2. Corporate Governance Variables
• Board size and Board of Directors_independence
3. Firm Characteristics variables
• Firm Size
• Firm Age.

In their observation, it was revealed that almost all the explanatory variables have significant and positive relationship with cyber-enabled and cyber-dependent crimes. In their study, they concluded that with the application of risk governance and other factors whose impact was tested and measured, cybercrime can be reduced, and impact minimized on the selected financial institutions that was used for the investigation [23].

"Characterising and Predicting Cyber-attacks using the Cyber Attacker Model Profile (CAMP)" is another piece of work whose methodology was heavily dependent and built on regression analysis. In their research, they formalised some of the methodologies to develop a cyber attacker model profile (CAMP) that can be utilised to distinguish and forecast cybercrimes [24]. The paper came up with a framework with the use of economic and social factors as the predictive or independent variables drawn from various countries in eastern Europe. Also included as the independent variables are yardstick or standard indicators of cyber related and cyber enabled crimes in the Australian financial service organizations. The study concluded that there is strong correlation existing between perceived corruption and the GDP in two different categories of nations. The outcome of the regression analysis formed an opinion that a mobile and highly skilled labour force in collaboration with a corruption enabling environment in countries under focus has great impact on and is greatly related to escalating cybercrimes.

From all these research works done so far on cybercrime profiling, it is obvious that cybercrimes are influenced by various factors. Investigations into these factors and measure of the impact they have on cybercrime will help curb this nefarious crime.

This project seeks to take a similar approach to analysing cybercrimes during the covit-19 pandemic. This is aimed at establishing the relationship between the lockdown caused because of the COVIT-19 pandemic and cybercrime during this period.

Further attempt will then be made to come up with a framework that will help project into the immediate future in the form of forecasting. This will serve as viable tool to enable various authorities, policy makers, heads of government parastatals and the entire stakeholders in the security sector to make key and accurate decisions on how to counter this nefarious act called cybercrime.

2.9 Hypothesis

After due consideration and detailed analysis of various works done so far within this very short period, the following hypothesis will be set.

H1: The COVID-19 pandemic and lockdown has immensely increased cyber dependent and cyber-enabled crimes.

H2: The Covid-19 pandemic and lockdown has not increased cyber dependent and cyber-enabled crimes.

3 Research Methodology

3.1 Introduction

Currently, the entire globe is in middle of a worldwide pandemic. With the current rule/law from the CDC for everyone to practice and maintain social distancing and avoid congregating or crowds of more than six people in a home, many system users, and members of staff of various organisations and firms have been mandated to work from home where possible. Various means have been introduced to help navigate through this new "normal". As we all must have come to terms with the fact that it has opened doors of opportunity for huge number of cybercrimes to be carried out.

In a rare time like this, cybercrime has taken a different dimension. So also, various other methods must be considered in combating this nefarious crime. In most cases of cybercrime profiling, which is one of the most effective way of exposing and fighting this crime from time immemorial, Routine Activity Theory- a qualitative method, has been adopted mostly. Cybercrime profiling and research industry is not left out- There is great need to look at cybercrime profiling using quantitative means and methods but with a different approach. This section deals with the research techniques used for the study. It is arranged under the following heading:

- Research Design
- Population of study
- Area of study
- Method of data collection
- Data analysis.

3.2 Research Design

Looking critically at this research works, it has some element of cross-sectional research/study, but it will be considered as a Longitudinal Research Design due to the following reasons.

- It is observational. This implies that the researcher has no influence whatsoever over the data. The data has been collected, collated and supplied by Hackmaggedon.
- It follows a time series dataset. Though it also follows a cross-sectional dataset i.e., a design that follows a time series dataset and cross-sectional dataset. A longitudinal research design is a data collected over a long period of time while cross sectional research design is the collection of data at a given point in time
- By virtue of scope and coverage, it investigates cause-and-effect relationships between attacks and the attack class than a cross-sectional study.
- Considering this study holistically, the research should drive or determine the design. But occasionally, the advancement of the research helps ascertain the most appropriate design. This research is not an exception to this fact. The research starts with a cross-sectional study to first establish whether there are relationships or associations between the variables. Then longitudinal study is now carried out to study cause and effect.
- This study/research will give room for the detection of development when the model is fitted. It will reveal developments or changes in the characteristics of the target class/population at both the group and the individual level. The key here is that longitudinal studies extend beyond a single moment in time. As a result, this will establish the sequences of cyber-attacks (theoretically).

3.3   Studied Population

The target population for this research will be the Cyber Crimes recorded during the core or main period of the COVIT-19 pandemic/lockdown from all over the world. This involves the cybercrimes committed, the target class and the attack class. Though the lockdown and pandemic are still on, but this piece of work aims to cover from January to September 2020 only.

The data is recorded cases of cybercrimes from all over the world within this period of the COVIT-19 pandemic and lockdown.

The method of data collection is a secondary data obtained from https://www.hackmageddon.com, a website that supplies information and Security Timelines and Statistics. The data were collected on monthly basis observe from January to September. 2020.

3.4   Limitations of the Data

The advent of the covid-19 pandemic has prompted the diversification of communication, businesses, and transactions all to be virtual. Scientists and technology so far have been on their toes working endlessly to see that while the economy, commerce and government does not crash, they are able to meet up with the challenges of this rare times to proffer solution to the crises caused by the pandemic. While all these elements of progress have been working for a positive paradigm shift, some person(s) also have been working tirelessly to hijack these privileges.

So also various efforts are being made by organisations like Hackmaggedon to gather data for researchers into this gruesome crime. But despite the commendable effort put into gathering these data, there are still some limitations.

Some of the demerits of the data collected includes.

1.   The responses are more of opinion responses (open ended questions) which is qualitative and for it to be relevant for this study it must be coded into close ended questions.
2.   Missing values (responses) on the authors (those bodies through which the crimes were perpetrated) involved in the criminal acts.
3.   Tracking the criminal acts as it was carried out in different counties
4.   Possibility of data being incomplete. There is the possibility of some cases not being reported for one reason or the other.

3.5   Data Analysis Methodology

3.5.1   Methodology

For every crime committed- be it traditional crime or cybercrime, various factors are always in play. In this case, this piece of work attempts to investigate the factors that influenced the increase in the number of crimes by studying the relationship between cybercrime occurrence and two major factors which are the attack class and the target class.

To achieve the aim and objectives of this research work, the following are proposed analysis that will be carried out, hopefully we will be able to measure the relationship between these factors and the number of cybercrimes recorded during the pandemic. Effort will then be made to forecast/project into the future and recommendations will be put forward on how to remedy the menace of cyber-crimes during and after the covid-19 pandemic period.

3.5.2   Descriptive Analysis

It will be good to recall that the data supplied by Hackmaggedon, though secondary, is still considered as a raw data. The Descriptive analysis forms the basis of every quantitative analysis. Here, the basic features of the data are being revealed. This will be done to give concise summaries about the sample and the measures. When descriptive analysis is being done on data, it presents the data in a manageable form in such a way that salient points can be deduced from the data. Descriptive analysis is also a good mathematical and/or statistical tools for easy, clear and accurate comparisons on a data to be analysed or used for a research. When applied on a data, comparison between two or different indices are been done in such a way that some or major differences, distinctions and similarities can be seen at a glance [25]. Though it does not give all that is to be known in a data, but it gives some very important information before the quantitative analysis is implemented full. Descriptive analysis helps to reveal character of a data.

It reveals the frequency of occurrence of a range of values for a particular variable. For instance, in the original data, it will be discovered that United State of America has been hit by cybercriminals over five hundred times this period.

A two-way Analysis of Variance (ANOVA) will be used to compare the relational differences of the measure of influences each of the explanatory or independent variables have on cyberattack occurrences during the COVIT-19 pandemic and lockdown covered by this period.

As part of the descriptive analysis, standard deviation will also be used. The standard deviation is one of the most detailed and accurate estimate of dispersion. The standard deviation will reveal the relation that the cybercrime occurrences have on the mean of the samples if we were to make inference.

For this piece of work, graphical and tabular representations will be adopted. Presentation of graphs and tables showing prevalence of cyber-crimes in the classification of various types of cyber-crime and so on.

3.5.3 Regression Analysis

When it comes to the study of relationships between two or more factors, one of the most viable and reliable tools for this is Regression analysis.

This study seeks to adopt the methodology Regression analysis of the predictor variables (attack class and attack target) on the dependent variable (cyber-crimes during the covid-19) to test the relationship between cybercrime occurrence and the predictor variables. This will be used to construct a model or framework in the form of $y = \beta x$ (see mathematical expressions in Sect. 3.6.1 below).

3.5.4 Time Series

Time series analysis of the data to test (and measure) the relationship between cyber-crime occurrence during the COVID-19 pandemic and time- t. Mathematically, it is stated as $y(t) = x(t)\beta + \varepsilon(t)$ (see expression in Sect. 3.6.2). It is also meant to forecast for the future occurrence and how to tackle the menace. One of the major selling points of this piece of work is the forecasting of cybercrimes to have a likely idea in terms of figures of the number of crimes that may be perpetrated in the nearest future.

1. Test for stochasticity will be carried out on the trend to see if it will be suitable for prediction or forecasting (see mathematical expression in Sect. 3.6.3 below)

3.6 Mathematical Model

A multiple regression model and time series model will be adopted for the purpose of predicting the cause and effect of the cyber-crime during and after the covid-19 pandemic period.

3.6.1 Multiple Regression Models

$$y = \beta_1 x_1 + \beta_2 x_2 + \cdots + \beta_n x_n$$

where
$\beta_{i's}(i = 1, 2, \ldots, n)$ is the regression coefficient which represents the value at which the criterion variables changes when predictor variable changes.
y = Cyber-crime rate.

x = Predictor or dependent variables (attack class and attack target).

3.6.2 Time Series Model

$y(t) = x(t)\beta + \varepsilon(t)$.

where; $y(t) = \{y_t; t = 0, \pm1, \pm2, \ldots\}$ is a sequence indexed by the time subscript t, which is a combination of an observable signal sequence $x(t) = \{x_t\}$ and an unobservable white-noise sequence $\varepsilon(t) = \{\varepsilon_t\}$ of independently and identically distributed random variables.

3.6.3 Test for Stochasticity of the Trend

After all descriptions, illustrations, and data representations above, one of the cogent and influential characteristics of the model or framework constructed and looking at the trend is that it is a continuous stochastic. This is since it is a collection of random (arbitrary and causal) variables indexed by time and it is a component of the attainment of the stochastic process.

Mathematically, assuming we have a probability space of triple expression as Q, V, P where:

Q represents a non-empty set referred to as the sample space.

V represents or symbolizes σ-algebra of the subset of Q, i.e. a group of the subsets closed with respect to a finite relationship and complement with regards to Q.

P is assumed to be the measure of probability defined for every or all subsets of V.

The real random (stochastic) variable on (Q, V, P) is a function x:Q → ℝ in such a way that the inverse image of any interval $[-\infty, a]$ belongs to V i.e. quantifiable function. The following terms if adopted, we can establish that the model is a stochastic process. The real random variable contains a family of real random variables $X = \{x_i(\omega); i\{\in U\}$, all defined on the same probability space (Q, V, P).

The set U refers to the index set of the process.

3.7 Statistical Package for the Data Analytics

The Statistical Packages for Social Science SPSS 20v will aid the researcher in carrying out the research analysis. The Descriptive Statistics using frequency count will help us identify the various cyber-crime commitment over the year, the country the target is, and probably the target. And the Chi-square analysis will help us identify the association between the cyber-crimes and other important variables in the data and finally the Time-Series analysis will help us to determine the nature and trend of the cyber-crime across the years with a suggested better model.

3.8   Data Coding

The data was coded using SPSS. Reasons include the following:

1. Repetition and redundancy: Since the data collected and entered contains some portion or measure of repetition and redundancy i.e., extra information that is not required for the analysis. This pattern or repetition has necessitated the coding of the data.
2. Speed of data entry: For entries like number of malware crimes committed, the entry will be done so many times. It will amount to exercise in futility since there is no additional information contained in the complete or full words when compared to a letter.
3. Increase accuracy of data entry: This is meant to reduce or eliminate the number of errors. Due to weariness and repetition of same word series of time, coding helped addressed this.
4. Use of validation: This ensures the ease of validation and authentication. By this possibility of error is reduced
5. Less storage space required: Since every letter when stored in the database, will require one byte of storage, to manage storage space, coding is very necessary. For a large data of cybercrimes recorded during the COVID-19 pandemic, large storage space will be utilised if not coded
6. Faster searching for data: Time taken to run queries or fetch data will not be much if data has been coded. The more reduced the size of a database, the faster and quicker will it be to comb and locate or fetch results.

4   Data Analysis

4.1   Introduction

In this section, effort is being made to cover the following major area which includes:

- Descriptive analysis of cybercrime during the COVIT-19 pandemic
- Estimation of Trend to discover behavioural patterns
- Making of Inference to trace down Causal effects
- Critical analysis of the result
- Predictive analysis of cybercrime occurrence based on current data using Time series and Regression Analysis
- Recommendations based on the findings.

4.2 Descriptive Analysis of Cybercrime During the Covit-19 Pandemic

4.2.1 Attack Class

This is the graph revealing the number of attacks carried out on these various targets (Tables 1 and 2) (Figs. 1 and 2).

Figure 3 and Table 3 above reveals the Attack class over the years. From the output, 89% (1164) attack CC via Cybercrimes Attacks, 10% (134) of the attack class were CE, 1.50% Attack Class were CW, 1.20% (18) attack class was H, finally 0.30% attack class were not revealed.

Table 4 shows that US has the highest Cybercrimes/Attacks/Attack over the years.

Table 1 Target class

Characteristics	Frequency	Percentage (%)
C Manufacturing	54	4.0
D Electricity gas steam and air conditioning supply	24	1.8
E Water supply, sewerage waste management, and remediation activities	2	0.1
G Wholesale and retail trade	46	3.4
H Transportation and storage	13	1.0
I Accommodation and food service activities	12	0.9
J Information and communication	36	2.7
K Financial and insurance activities	90	6.7
L Real estate activities	4	0.3
M Professional scientific and technical activities	74	5.5
N Administrative and support service activities	5	0.4
O Public administration and defense, compulsory social security	159	11.9
P Education	110	8.2
Q Human health and social work activities	134	10.0
R Arts entertainment and recreation	35	2.6
S Other service activities	26	1.9
U Activities of extraterritorial organizations and bodies	11	0.8
V Fintech	23	1.7
X Individual	202	15.1
Y Multiple Industries	262	19.6
Z Unknown	16	1.2
Total	1338	100.0

Sources Researcher's SPSS Output, 2020s

Table 2 Attack/Cybercrimes

Characteristics	Frequency	Percentage (%)
Account hijacking via QR-code	223	16.7
Android vulnerability (CVE-2020–0032)	1	0.1
API Exploit	1	0.1
AWS misconfiguration	2	0.1
BGP hijacking	1	0.1
Bitcoin vanity addresses	1	0.1
Brute-Force	5	0.4
Business email compromise	16	1.2
Citrix vulnerability (CVE-2019-19,781)	3	0.2
Cloud misconfiguration	6	0.4
CMS vulnerability	1	0.1
Credential stuffing	6	0.4
CVE	26	1.9
DDoS	35	2.6
Defacement	9	0.7
DNS hijacking	2	0.1
Domain spoofing	1	0.1
DrayTek Vigor enterprise routers vulnerability	1	0.1
Facebook vulnerability	1	0.1
Fake social network accounts/pages	12	0.9
Ghostcat (CVE-2020–1938) vulnerability	1	0.1
Google Chrome vulnerability (CVE-2020–6418)	1	0.1
Lilin vulnerabilities	1	0.1
Malicious bot	1	0.1
Malicious browser extension	4	0.3
Malicious npm package	1	0.1
Malicious script injection	41	3.1
Malicious spam	30	2.2
Malicious wordpress plugin	2	0.1
Malicous autodesk plugin	2	0.1
Malvertising	5	0.4
Malware	532	39.8
Microsoft vulnerability (CVE-2020–0688)	1	0.1
Misconfiguration	13	1.0
Misconfigured AWS S3 Bucket	1	0.1
Misconfigured citrix devices	1	0.1

(continued)

Table 2 (continued)

Characteristics	Frequency	Percentage (%)
Multiple vulnerabilities	1	0.1
Password-spray	3	0.2
Path traversal vulnerability	1	0.1
PayPal vulnerability	1	0.1
PoS malware	4	0.3
QNAP vulnerability	1	0.1
Router vulnerability	1	0.1
Salt vulnerability	5	0.4
Search engine poisoning	1	0.1
Server-side ad insertion (SSAI) hijacking	1	0.1
Shellshock vulnerability	1	0.1
SQL injection	9	0.7
SS7 vulnerabilities	1	0.1
Symantec web gateway vulnerability	1	0.1
Targeted attack	126	9.4
Trend micro vulnerabilities	1	0.1
Unknown	161	12.0
Vulnerability	7	0.5
Vulnerable wordpress plugins	10	0.7
Web shells	1	0.1
Wiretapping	1	0.1
Zero-day 54	1	0.1
Zoom bombing	8	0.6
Zoom 34	1	0.1

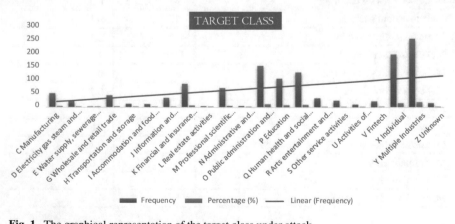

Fig. 1 The graphical representation of the target class under attack

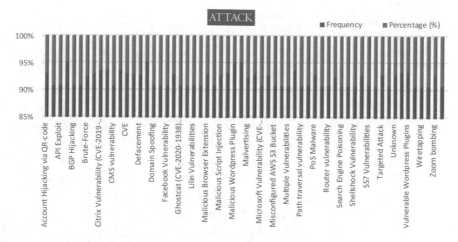

Fig. 2 The attack across the year

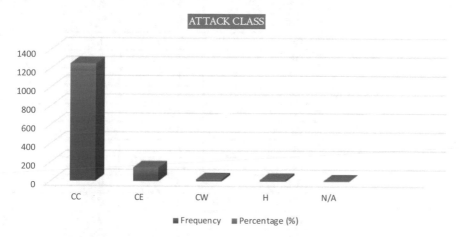

Fig. 3 The attack class across the year

Table 3 Attack class

Characteristics	Frequency	Percentage (%)	MEAN	SD
CC	1164	87.00	30.02	0.509
CE	134	10.00	28.88	1.465
CW	20	1.50	25.85	2.971
H	16	1.20	28.50	4.815
N/A	4	0.30	24.50	10.882

Sources Researchers' SPSS Output (2020)

Table 4 The country under attack across the year

Characteristics	Frequency	Percentage (%)
AM	2	0.1
AR	1	0.1
AT	2	0.1
AU	25	1.9
AZ	2	0.1
BH	1	0.1
BR	7	0.5
BY	2	0.1
CA	28	2.1
CH	6	0.4
CL	1	0.1
CN	9	0.7
CR	2	0.1
CW	2	0.1
DE	20	1.5
DK	3	0.2
DZ	1	0.1
EE	1	0.1
EG	1	0.1
ES	10	0.7
EU	3	0.2
FR	14	1.0
GE	2	0.1
GR	2	0.1
HK	2	0.1
HR	1	0.1
HU	2	0.1
ID	7	0.5
IE	3	0.2
IL	8	0.6
IN	22	1.6
INT	1	0.1
IR	5	0.4
IT	17	1.3
JM	1	0.1
JP	12	0.9
KR	11	0.8

(continued)

Table 4 (continued)

Characteristics	Frequency	Percentage (%)
KW	1	0.1
LB	1	0.1
LK	1	0.1
LY	1	0.1
MA	2	0.1
MN	1	0.1
MT	2	0.1
MX	2	0.1
MY	1	0.1
N/A	30	2.2
NL	5	0.4
NO	2	0.1
NP	2	0.1
NZ	5	0.4
OM	1	0.1
PE	1	0.1
PH	5	0.4
PK	6	0.4
PL	1	0.1
PR	1	0.1
PS	2	0.1
PT	6	0.4
RS	1	0.1
RU	8	0.6
RW	2	0.1
SA	2	0.1
SC	2	0.1
SE	1	0.1
SG	8	0.6
SK	3	0.2
SL	2	0.1
SY	1	0.1
TH	1	0.1
TR	3	0.2
TW	3	0.2
UA	7	0.5
UK	52	3.9

(continued)

Table 4 (continued)

Characteristics	Frequency	Percentage (%)
UNKNOWN	412	30.8
US	501	37.4
VA	1	0.1
VN	3	0.2
ZA	5	0.4
Total	1338	100.0

Table 5 Trend of cybercrime over the year

Months	Attack (%)
Jan	11.8
Feb	14.4
Mar	14.0
Apr	14.1
May	8.7
Jun	13.4
Jul	0.1
Aug	8.5
Sep	15.0
Grand total	100.00

Sources Excel Pivot Table 2020 Output

The graph and the Table 5 above shows that the nature of the crime committed across the year follows a non-stationary model i.e. the mean and the variance vary across the months. All through the period of the pandemic across the month, looking at the trend it follows,

4.3 Critical Analysis

From the Fig. 1 and Table 1, the graphical representation and tabulation of target class of the Cybercrimes across this period was determined.

Looking at the number of attacks and the target class, it was spread across almost all sectors. But how ever some sectors were more frequent than others. Let us critically examine the top three targets that recorded the highest as seen in this data during the COVIT-19 pandemic and lockdown as there are some information that will help in cybercrime profiling in a critical period like this.

The target class that experienced the highest series of attack is the one referred to as "multiple Industries" with over two hundred and sixty attacks and forms over nineteen percent of the total attack. Though this signifies a sort of a compound name or like a group comprising of different set of people. A closer look at the original data reveals that the multiple industries as stated there comprises the following set of people:

- The vulnerable users
- Ruby Users (Ruby is a flexible object-oriented programming language [26])
- Organizations
- Energy companies [27]
- Manufacturing
- Zoom users
- Multiple users like religious organizations, group of friends, family members, social clubs, association of friends and so on.

The second target class with the highest hit are those referred to as "individuals" with about a total of two hundred and two attacks. The third is another group with two different sectors joined together- The Public administration, defence, and social security. This forms about over eleven percent of the total attack.

Furthermore, Fig. 2 and Table 2 reveals the type of attacks carried out on the different sectors over this specific period of COVIT-19 lockdown in year 2020.

The highest type of attack or attack class is Malware attack with a total of about five hundred and thirty-two attacks with approximately forty percent of the total attacks recorded in this data. Roy Reynolds in his research works established that malware attack is one of the greatest cyber-threat to business establishments in the United Kingdom and even Great Britain as a whole [28]. Recent series of research and records of Malware attack reveals that it is always a part of a much bigger attack or hit [29]. Malware is used to gain control to a victim's privacy (valuable documents and files) so as to carry out the real intended act of crime [30].

Account Hijacking via QR-code is the second most used type of attack as recorded during this period that is referred to as a new-normal. An Egyptian researcher in the information security sector and who is also a cyber security consultant with Seekurity incorporated came up with a proof-of-concept illustrating a modern session of hijacking methods that can be used to hijack accounts from service users that log in with the QR code (Quick Response Code). It was named QRLJacking. This type of attack or attack class is a simple attack vector that influence or induce all the applications that require login with the QR code or feature. [31]. The diagram below explains how this is being done. All the attacker need is to find a way to convince the unsuspecting victim to scan the attacker's QR code.

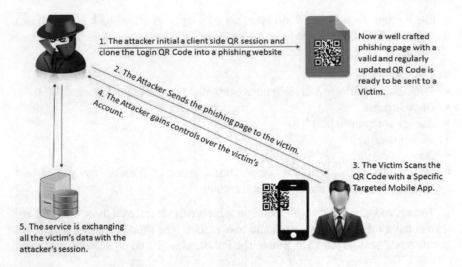

QRLJacking Technique [31]

Other attacks according to the frequency of attack on the data include malicious Script Injection, malicious spam, Distributed Denial of Service (DDOS), Common Vulnerabilities and Exposure (CVE) and others.

Again, a critical look at the various countries attacked, revealed that United State of America suffers the highest number of attacks (see Fig. 4 and Table 4). Australia, United Kingdom, and some other countries are not left out.

Figure 5 and Table 5 reveals the distribution of the nefarious crimes being spread across the period from January to September 2020. This takes us to the trend analysis.

The result in Table 6 above shows the regression model summary. From the output, the correlation coefficient [R = 0.926] implies that there is a strong positive correlation between the dependent variable (Attack/Cybercrimes/Attacks/Attacks) and the explanatory variables (Attack Class, Target Class) as explained in the table.

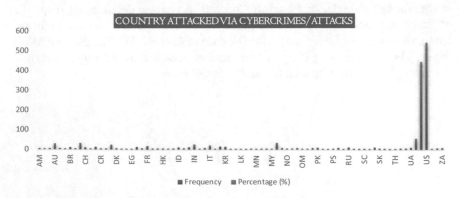

Fig. 4 Country attack via cybercrime attacks

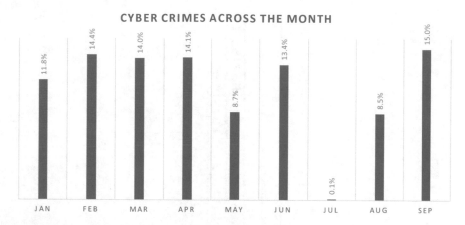

Fig. 5 Trend of cybercrimes/attack/attack across the month

The coefficient of determination [$R^2 = 0.858$] dictate that 85.8% of the variation in the dependent (Attack/Cybercrimes/Attacks/Attacks) variable can be explained by the explanatory variable (Attack Class, Target Class). While 14.2% can be explain by other factors in the model and the (Adjusted $R^2 = 0.858$) shows the adequacy of the model.

Table 7 reveals the regression analysis model which is given as:

$GDP = -10.405 + 2.858$ Target Class $+ 0.772$ Attack Class.

Furthermore, the model shows a strong positive linear regression model which illustrates that the explanatory variables-Attack Class and Target Class are good determinant factors of the dependent variables (Cybercrimes Attacks) and that a unit increase in Target Class will contribute 2.858 increase in Attack/Cybercrimes/Attacks/Attack keeping Attack Class constant by 10.405 and a unit increase in Attack Class will yield 0.772 increase in attack/Cybercrimes/Attacks/Attacks keeping Target Class constant by 10.405. The sig value [(0.000 & 0.020) < 0.05] shows that the explanatory variables to the dependent variables are statistically significant.

The test for multicollinearity was determined. Multicollinearity occurs when independent variables in a regression model are correlated.

Looking at this test, special consideration was given to the p-value. Then, multicollinearity was estimated knowing fully that it will be an issue that is peculiar in the analyses of models fitted from multiple regression if it is too high. To ensure that the multicollinearity is brought to the barest or minimum level, first some simple correlations were made between pairs of independent variables before setting out the layout of the multiple regression.

The Variance Inflation Factor (VIF) helps the determination of presence of multicollinearity in a variable. If the VIF $= 1$ it means that there is no correlation between the independent variables but if the VIF is between 1–5 then there is a moderate correlation, but it is not severe enough to warrant corrective measures and the VIF > 5 represent critical levels of multicollinearity where the coefficients are poorly

Table 6 Regression model summary

Model summary

Model	R	R square	Adjusted R square	Std. error of the estimate	Change statistics				
					R square change	F change	df1	df2	Sig. F change
1	0.926[a]	0.858	0.858	6.515	0.858	4041.670	2	1335	0.000

[a]Predictors: (Constant), Attack Class, Target Class
[b]Dependent Variable: Attack

Table 7 Regression coefficients and multicollinearity test coefficients[a]

Model		Unstandardized coefficients		Standardized coefficients	t	Sig.	Correlations			Collinearity statistics	
		B	Std. error	Beta			Zero-order	Partial	Part	Tolerance	VIF
1	(Constant)	−10.405	0.633		−16.437	0.000					
	Target Class	2.858	0.032	0.928	89.843	0.000	0.926	0.926	0.926	0.996	1.004
	Attack Class	0.772	0.335	00.024	2.305	0.021	−0.035	0.063	0.024	0.996	1.004

[a]Dependent Variable: Attack

estimated, and the p-value are questionable. So, from the table we can depicts that the VIF is between 1–5 and we can say that there is a moderate correlation among the independent variables. A look at the VIF of the target class and the attack class gives us 1.004 which indicates that though there is a mild, minute or a moderate correlation, it is not enough to pose any threat and does not warrant or call for any corrective measure to be taken (Table 8).

The table reveals the ANOVA (Analysis of Variance) result of the variables concerned. From the output Sum of Square Regression and Residual respectively are (6135824733.519 *and* 1290941171.670). df = 3, 29, and the mean difference respectively are (2045274911.173 *and* 44515212.816) and the p-value(sig = 0.000). Since the significant (p-value = 0.00 < 0.05), then depict that explanatory variables [Attack Class, Target Class] are statistically different meaning that they are key factors contributing to Cybercrimes Attacks and if measure is not taking the cybercrime will significantly increase.

Obviously, a particular factor or event is at play here because looking at the rate at which cybercrime occurrences is escalating. This establishes the influence or impact of the lockdown as a result of the pandemic. Research reveals or that has shown that anytime there is a major event, there is always a significant influence in crime. This influence can be to reduce it or to make it increase [18].

When it comes to traditional crimes, it will be seen clearly that the rate of occurrence dropped drastically during this lockdown [18].

But in the case of cybercrime, the reverse is the case.

From this, inference can be made that due to restriction of movement as a result of the lockdown and the fear of getting infected, traditional crimes drastically reduced.

Going further, the idea of working from home adopted by most organisations has diverted the traffic of crime to the web. Due to the lockdown, most transactions, discharge of official duties, contacting of friends and family members, social gatherings, lectures, religious gatherings, buying and selling and so on have all been moved to online and hence the paradigm shift in crime from traditional crime to cybercrimes. This explains the rise in cybercrime during this period.

Also, due to the fear of the pandemic, the distress it created, some are depressed, some are bored, some worried, roads deserted, streets empty, and everyone is forced to sit at home, it can also be deduced that the alert levels of most users have dropped thereby making them victims of this nefarious act. In some houses, even under same

Table 8 ANOVA test

ANOVA[a]						
Model		Sum of squares	df	Mean square	F	Sig
1	Regression	343056.989	2	171528.494	4041.670	0.000[b]
	Residual	56657.413	1335	42.440		
	Total	399714.401	1337			

[a]Dependent Variable: Attack
[b]Predictors: (Constant), Attack Class, Target Class

roof, due to idea to isolate, members of family cannot even relate with each other. All these issues contributed immensely to the increase in cybercrime.

The second most common method of cyberattack according to the data from Hackmaggedon used for this analysis- Account Hijacking via QR-code, it will be clearly seen that most users acted out of depression, boredom, diuresis, worry, anxiety and so on. In conclusion, most of the users acted with an unstable mind caused as a result of the lockdown.

More worrisome is the fact that the figures of the cybercrime occurrences are really increasing, if drastic measures are not taken, there will be serious and devastating consequences on the web environment which is fast becoming the arena of life and activities due to its universal embrace.

At this junction, it will be good to make some projections and forecast into the future of cybercrime occurrences if this lockdown persists.

4.4 Time Series Analysis and Forecasting of Cybercrimes

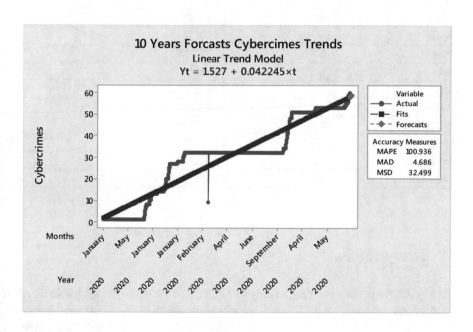

2021–2030 Yearly Forecasts (10 years)

Year	Forecast
2021	58.0922

(continued)

(continued)

Year	Forecast
2022	58.1345
2023	58.1767
2024	58.2190
2025	58.2612
2026	58.3035
2027	58.3457
2028	58.3879
2029	58.4302
2030	58.4724

2021 monthly forecasts of cybercrimes

Months	t	$Y_t = 1.527 + 0.042245 \times t$
January	1	1.569245
February	2	1.61149
March	3	1.653735
April	4	1.69598
May	5	1.738225
June	6	1.78047
July	7	1.822715
August	8	1.86496
September	9	1.907205
October	10	1.94945
November	11	1.991695
December	12	2.03394

4.5 Interpretation

The Fig Above reveals the Trend analysis (2020) and 10 years (2030) forecasts of Cybercrimes across the years. From the results above, there has been an upward trend over the years meaning that Cybercrimes increases as the months increases. Also, the trend has been a non-stationary series i.e. the mean and the variance changes across time.

The 10 years forecasts give the estimated model $Y_t = 1.527 + 0.042245 \times t$ which is a strong determinant of the cybercrimes.

Meaning that a unit increase in Time will yield 0.042245 increase in cybercrimes crimes in order words, an increase of 0.042245 every year will increase the value of cybercrimes. Therefore, the table above reveals a future increase in the value of Cybercrimes.

4.6 Recommendations and Conclusion

Looking at this piece of work so far, drastic steps must be carried very fast as the indicators clearly reflects danger if it continues like this. A critical look at all the indicators in this piece of work reveals that a major event has contributed to the spike in cybercrime occurrences. No doubt, the only event currently is the lockdown because of the COVIT-19 pandemic.

One very clear and obvious cause is that many system users are forced to work in a technological environment that lacks enough protection against such crimes. One very unfortunate fact is that most of these users are ignorant of the devices of these cyberfraudsters.

A closer look at the highest type of cybercrime used during this pandemic reveals is Malware attack. As we know, a malware attack is a situation whereby a cybercriminal produces a malicious application that is installed on the victim's system with his or her knowledge to create an access or a backdoor into the victim's data.

Looking at the prevalent methods or mode of attack during the lockdown introduced by CDC and WHO to alleviate or reduce the spread of this this deadly pandemic, they are targeted at mostly individuals and members of staff of large and small organisations who have some special elevated right and privileges on the network to hit their target.

The following recommendations will go a long way in reducing cybercrimes during in a time like this.

5 Conclusion and Future Work

Despite that this piece of work has made several significant and symbolic discoveries and findings, there are still lots of work to be done and a lot of grounds to cover.

One of the major outstanding work to be done is in individual vulnerability to cybercrime due to limitations in methodological capabilities. Though looking at it holistically, major factors that gave enabling environment to cybercrime during this period has been established, but there is great need in the future to address individual lapses and area of vulnerabilities by profiling cybercrime to show specific and individual patterns and character of various actors and component of a cybercrime cycle (the cybercriminal, the victim, and the vulnerability).

Another very important area to also work on is data collection. No doubt about it, Hackmaggedon is known and commended for good work when it comes to collection of data and information for the purpose of research works and studies in the field of cyber security, but some information or data are not captured. This data was supposed to be cybercrime occurrences during this period for all over the world, but there are lots of thousands of individuals that were hacked but not captured in data collection. There were series of cybercrime all over the world that was not captured. Another issue with the data again is that some of them were open-ended with the researcher left to supply the missing link. In the future, it will be good if research works can beam its searchlight on data collection. This also goes to world government, establishments, and all stakeholders in the cyber security industry.

It is also believed that if a complex multi-faceted research work is done to harmonise previous works and harness their potentials to investigate cybercrime, it will go a long way in curbing cybercrime.

Irrespective of the shortfalls or limitations of this research work, it has been able to explore situational factors and its influence on cybercrimes during the lockdown caused by the pandemic. It has also further revealed to an appreciable and a recognisable extent that cybercriminals exploits situational factors to hit their targets.

One very important area that still requires great works of research to be done in the future is the unemployment rate. One of the implications of this lockdown is unemployment. As a result of the lockdown and sit-at-home enforcement, more people are online, idle and are at home [11]. Looking at this, there is a high probability that most of the will likely begin to explore cyber opportunities and become a cybercriminal to put food on the table and/or keep themselves busy. This is another very important research area to explore in future works.

However, future research works could engage experimental research models and framework to profile cybercrime in a virtual social or economic crisis like the one created by this pandemic.

It is my opinion as a researcher that great research works like this should not be done during the crises. It would have been better done during a relaxed period but in a simulated environment.

Hopefully, as more investigations and studies on cybercrime are being done and more countermeasures are being employed, the world wide web will be safer and free of all these criminal acts for all users and entire stakeholders in cyber world.

References

1. Business CNN (2020) Stocks sink as US oil prices fall below $0 a barrel: April 20, 2020. CNN Business, USA
2. Macdonald K (2019) Cybercrime: awareness, prevention and response. s.n., Canada
3. WHO (2011) The classical definition of a pandemic is not elusive. s.n., Australia
4. Dan L (2020) Ransomware during Covid-19. Lohrmann on cybersecurity and infrastructure
5. Levi M (2016) Assessing the trends, scale and nature of economic cybercrimes: review and issues. Springler Link 67(February 2017):3–20
6. Hassold C (2019) The threat taxonomy: a working framework to describe cyber attacks [Online]. https://www.agari.com/email-security-blog/threat-taxonomy-framework-cyber-attacks/. Accessed 12 Oct 2020
7. Yar M, Leukfeldt ER (2016) Applying routine activity theory to cybercrime: a theoretical and empirical analysis. Defiant Behav 37(3)

8. CPS (2019) CPS [Online]. https://www.cps.gov.uk/legal-guidance/cybercrime-prosecution-guidance#:~:text=a%20cybercrime%20case.-,Definitions,distinct%20ranges%20of%20criminal%20activity.&text=Cyber-enabled%20crimes%20-%20traditional%20crimes,enabled%20fraud%20and%20data%20theft. Accessed 26 Oct 2020

9. Chandraa A, Snowe MJ (2020) Taxonomy of cybercrime: theory and design. Science Direct 38(100467)

10. Singh MM, Bakar AA (2019) A systemic cybercrime stakeholders architectural model. Elsevier ScienceDirect (161):1147–1155

11. Lallie HS, Shepherd LA, Nurse JRC, Erola A, Epiphaniou G, Maple C, Bellekens X (2020) Cyber security in the age of COVID-19: a timeline and analysis of cyber-crime and cyber-attacks during the pandemic. arxiv.org

12. Asmir Butkovic SMSUAT (2018) Geographic profiling for serial cybercrime investigation. Elsevier 28:176–182

13. Kim H-W, An J (2018) A data analytics approach to the cybercrime underground economy. IEEE Access 6:1–17

14. Evans D (2020) How Zoom became so popular during social distancing [Online]. https://www.cnbc.com/2020/04/03/how-zoom-rose-to-the-top-during-the-coronavirus-pandemic.html. Accessed 19 Oct 2020

15. Hawdon J, Parti K, Dearden TE (2020) Cybercrime in America amid COVID-19: the initial results from a natural experiment. Am J Crim Justice 45:1–17

16. Mikolai ER (2020) Cybercriminals exploitation of the coronavirus pandemic

17. Ibrahim S (2016) Social and contextual taxonomy of cybercrime: socioeconomic theory of Nigerian cybercriminals. Int J Law Crime Justice 47:44–57 (1756–0616/©2016)

18. Felson M, Stickle B (2020) Crime rates in a pandemic: the largest criminological experiment in history. Am J Crim Justice 45:525–536

19. Morrow J (2020) Routine actvity theory. s.n., San Francisico

20. Kitteringham G (2010) Environmental crime control. Science Direct 151–160

21. Collins H (2013) Predicting crime using analytics and big data. Government Technology (GT), Chicago

22. Buil-Gil D, Miró-Llinares F, Moneva A, Kemp S, Díaz-Castaño N (2020) Cybercrime and shifts in opportunities during COVID-19: a preliminary analysis in the UK. Eur Soc 1–14. https://www.tandfonline.com/loi/reus20

23. Erin OA, Kolawole AD, Noah AO (2020) Risk governance and cybercrime: the heirarchical regression approach. SpringerOpen J (Fut Bus J) 6(12):1–15

24. Watters PA, McCombie S, Layton R, Pieprzyk J (2012) Characterising and predicting cyber attacks using the cyber attacker model profile (CAMP). J Money Laund Control 15(4):430–441

25. Stone H, Thomas HA (2020) Chapter 6—descriptive analysis. ScienceDirect 235–295

26. Andress J, Linn R (2017) Chapter 4—introduction to ruby. In: Coding for penetration testers, 2nd edn. s.l., Elsevier, pp 111–149

27. Naidoo R (2020) A multi-level influence model of COVID-19 themed cybercrime. Eur J Inf Syst 29(3):306–321

28. Reynolds R (2020) The four biggest malware threats to UK businesses. Elsevier/Sciencedirect 2020(3):6–8

29. Ngo Q-D, Nguyen H-T, Le V-H, Nguyen D-H (2020) A survey of IoT malware and detection methods based on static features. Sciencedirect 6(4):280–286

30. Soltys M, Gittinsa Z (2020) Malware persistence mechanisms. Elsevier/Sciencedirect 176:88–97

31. Khandelwal S (2016) Hacker news [Online]. https://thehackernews.com/2016/07/qrljacking-hacking-qr-code.html. Accessed 05 Jan 2021

The Role of Deep Neural Network in the Detection of Malware and APTs

Tr Thambi-Rajah and Hamid Jahankhani

Abstract Malware is a huge category of software, generally designed with malicious intent. It has gained much attention in the media over last decade with a number of high-profile ransomware cases coming to light. Malware is often used as part of a kill chain in Advanced Persistent Threats (APTs) attacks. It can easily be developed using tools such as "Veil Framework" by armature malware writers, or created from scratch by highly skilled individuals or teams as part in organised crime, or even by foreign agencies as part of an advanced cyber intelligence programme. As malware continue to evolve, at an ever-increasing rate the security community has turned to machine learning in a quest to find an effective method for detect and identify malware quickly. In response modern malware, is using more sophisticated detection-avoidance methodologies, against both binary analysis, and dynamic analysis. Thus, to preform, traditional static analysis requires advanced tool and an ever-increasing level of skill. Exploration of these various detection avoidance techniques as been included for understand of the issues encountered by security professionals, along with how improvements in software development methodologies and tooling has helped malware authors avoid detection, for little or no effort. This chapter explores the development of Deep Neural Networks in the defence against malware, looking specifically at how recent advances in Image Classification can be used detect and classify malware. As the Machine Learning industry seeks to make artificial intelligence and machine learning more accessible, through zero-code deep learning model training, this chapter will explore whether a zero-code approach to machine learning is feasible for developing deep neural network models for detecting and classifying malware. In particular the use of image classification models shall be explored, using off-the-self free or low-cost software, or cloud services.

Keywords ByteClass · Machine learning · Google AutoML · APT · Deep neural networks · Malware · Zero-code deep learning · AI · Artificial intelligence

T. Thambi-Rajah · H. Jahankhani (✉)
Northumbria University, London, UK
e-mail: Hamid.jahankhani@northumbria.ac.uk

© The Author(s), under exclusive license to Springer Nature Switzerland AG 2021
R. Montasari et al. (eds.), *Challenges in the IoT and Smart Environments*,
Advanced Sciences and Technologies for Security Applications,
https://doi.org/10.1007/978-3-030-87166-6_7

1   Introduction

1.1   Aims

Much research has been invested in Deep Learning for Cyber Security, advanced Deep Neural Networks and Convolutional Neural Networks (a sub-branch of Deep Neural Networks) have been created for the sole purpose of malware detection, and classification. However, with recent advances in Artificial Intelligence and many off-the-shelf Machine Learning Computer Vision application now available, would it be possible to create a Deep Neural Network without any coding?

1.2   Background

Ideally Malware would be detected and blocked at the Network Parameter, and while much work is happening on this front it is proving not to be enough to prevent malware infections on target devices be them IoT, PCs, Server, Mobile Devices or other. On the other hand, Dynamic Analysis of Malware is the final line of defence, allowing identification of Malware once the infection has occurred. Thus, making Static Analysis vital to any Defence in Depth approach against Malware. Beyond this, Static Analysis is well positioned for detecting Malware in both transit and storage, before execution. For instance, it can be used to analysis file on remote storage, in emails, on CDs and much more. With an ever-increasing sophistication in malware design there are circumstances where Dynamic Analysis leads to less than stratifying results, especially where the malware is able to detect its analysis environment and cloak its behaviour, or where the malware requires particular constraints to be met or an external trigger to occur to trigger its malicious behaviour [1].

2   Background

The definition of an APT generally consist of an adversary possessing advanced expertise or a sophisticated skillset allowing them to utilise multiple attack vectors, to achieve an objective against an organisation [2].

A study on "Advance Persistent Threats" rightly assets, that APT are generally distinguished by:

1. Specific target and clear objectives
2. Highly organized and well-resourced attackers
3. A long-term campaign with repeated attempts
4. Stealthy and evasive attack techniques.

This is consistent with the definition provided by "*Feature analysis for data-driver APT-related malware discrimination*", which expands on this stating that motivation for an APT are generally for business or political reason, however the article then attempts to qualify the use of the word "persistent" in the APT acronym giving it the definition as "*continues being undetected, being monitored and data being extracted for a long period of time*". One could argue that Ransomware only need hide its existence for a short period of time, to encrypt user-data before making itself known to the user, and demanding the ransom. Thus, this explanation does not entirely hold true, as APTs has moved beyond the realm of government and financial organisations, into all industries such as Healthcare, Education and Software Development and Media, etc.

A generalised assumption has been made in the industry that APT are carried out by well resourced, and organised teams. However, it would not be inconceivable that a highly skilled assailant to orchestrated an APT as an insider, and depending on the target organisation the number of resources required may be reduced. Thus, we should not assume that all APTs are executed by teams.

Given the differences between APTs and traditional attacks as outlined by Chen in his "Comparison of traditional and APT attacks" (below). The most defining trait of an APT is the approach, where an attacker is not easily dissuaded from attempting to compromise a system, and will continue to attempt to attack a system for an extended period of time, until they are able to breach a system (Table 1).

Much work has been carried out on the identification of malware, by analysing network traffic, static analysis and behaviour analysis. However few pieces of work specifically address the detection of malware belonging to APTs. This may be due to a number of reasons:

1. The defending organisation may not want to let the attacker know that they are aware of the attack. As in the case of Honey Pot environments, where the defending team will create environments to attract attackers, so that they can understand the techniques used, and develop defences for their production environments.
2. The defending organisation may not want to publicly acknowledge that they have been compromised, as this may carry reputational risk.

Table 1 Comparison of traditional and APT attacks [2]

	Traditional attacks	APT attacks
Attacker	Mostly single person	Highly organized, sophisticated, determined and well-resourced group
Target	Unspecified, mostly individual systems	Specific organizations, governmental institutions, commercial enterprises
Purpose	Financial benefits, demonstrating abilities	Competitive advantages, strategic benefits
Approach	Single-run, "smash and grab", short period	Repeated attempts, stays low and slow, adapts to resist defences', long term

3. APT attacks are specific to each target, as such there is limited commercial
    viability in investing research into any specific defences for specific APTs.

2.1 Malware

The Collins English Dictionary defines Malware as:

> **Malware**('*maepwea) n software that is designed to specifically to damage or disrupt a
> computer system [C20: from MAL(ICIOUS) + (SOFT)WARE].* [3]

Within the Technology industry malware is recognised as an all-encompassing
term for malicious software that includes Adware, Cryptojacking, Ransomware,
Rootkits and Spyware.

Well known victims of Malware, include:

- Nation Health Service—WannaCry 2017
- Northumbria University—Ransomware Doppelpaymer gang [4]

Above are well-known attacks that are well documented and were made public
either due regulatory requirements, or large-scale impact which necessitate a public
announcement. However, gaining accurate statistics on the use of malware and the
ever presence of APTs is difficult, as known attacks often go unreported, and there
are further number of undetected attacks.

It is useful to use a high-level categorisation of computer files noted by a
"Comparative Analysis of Malware Detection Techniques" [5]:

- Benign Files—Files do not cause any harm to a system or user.
- Normal Files—Files do not contain any form of malware.
- Grayware—Contains annoying, undisclosed and undesirable behaviour.
- Malware—File that cause harm to a system or user (Fig. 1).

Fig. 1 Types of files in
computer/cyber security

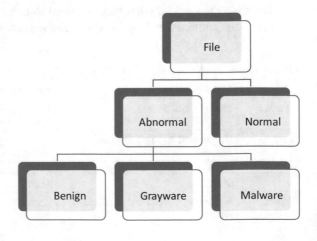

As a result, organisations are looking towards next generation technologies to detect and defend against Malware attacks. Many have invested heavily in the use of Artificial Intelligence (AI), Anomaly Detection, and Fingerprinting to detect the presence of Malware, with varying success.

2.2 Deep Learning

Deep learning (DL) is a branch of Artificial Intelligence (AI). Fundamentally (DL) is category of Neural Networks with multiple hidden layers of Neural Networks (NN). Deep Learning can be further categorised by the type of Neural Networks that they utilise, which includes but it not limited to:

- Convolutional Neural Networks (CNN)
- Deep Belief Networks (DBN)
- Generative Adversarial Networks (GAN)
- Networks (GAN)
- Recurrent Neural Networks (RNN)
- Restricted Boltzmann Machines (RBM).
- Autoencoders.

With break through in Deep Leaning techniques and advances in parallel computer, we have seen a huge increase in the use of AI in everyday applications. Individuals, companies and criminals, have taken keen interest in Artificial intelligence and Deep Learning.

2.2.1 Individuals

Individuals are taking advantage of the latest leaps in AI though common task automation using AI. But also, in less noticeable ways such as Optimised Battery Charging, as introduced in iOS13 which monitors a user's phone usage habit's and then sets the optimum pattern to compete the mobiles phone battery recharge cycle.

2.2.2 Organisations

Organisation have used Deep Learning for many years in areas such as Time Series Analysis, Data Mining, and Quality control. However, as the technology has become more accessible and cheaper to implement, organisations are constantly on the lookout for where else they can employ AI. Adobe has implemented AI for intelligent image selection in 2D photographs, Teslar has implemented AI to enable Autonomous Vehicles, while Cisco has enabled AI for Endpoint Security.

2.2.3 Criminals

Criminals are interested in the various opportunities that AI has presented to commit not only Cyber Crime, but Social Engineering. For instance, GANs have been used successfully to generate Deep Fakes. Deep Learning is not only able to change the voice of one individual to sound like another individual, but can even allow an individual to generate a video message using the face of another person. As demonstrated by the Obama Jordan Peele Deep Fake (Table 2).

Machine Learning has become such a critical part of today's computing experience, the Apple has integrated a Neural Engine on their latest chips (M1 for personal compute, and A14 for mobile compute). While Nvidia continues to invest into the CUDA-X Ai Libraries which harnesses the vast number of simple cores available on the Nvidia CPU for Deep Neural Networks (https://docs.nvidia.com/deeplearning/cudnn/index.html).

While it has been easy for organisation to identify, previously automated task and assess their suitability for enhancement with Ai, it has not always been easy to image new uses for Ai, nor has is it always easy to identify the most appropriate category of Ai to implement.

As expected, this massive uptake in AI has not gone unnoticed in the Cyber Security Industry, and many organisations have enabled Ai for both Endpoint Security and Network Surveillance. There has been much research into the use of Deep Leaning for traffic analysis and application behaviour analysis.

Table 2 Deep fake video—Jordan Peele as Obama [6]

3 Literature Review

Since the introduction of anti-virus programs, malware writers have been devising various ways to evade them. In response security specialist have developed and introduced various methods to detect and remove malware from infected systems.

3.1 Patterns of Malware Detection Avoidance

There are various methods of malware detection. A common anti-virus detection method is binary signature-based detection, this method simply searches for known binary signatures and flags files that contain that that contain these signatures as being infected. This simplistic methodology relies on a precompiled databased of known malware signatures, which may fail to detect the presence of known malware where the malware is polymorphic or metamorphic. Furthermore, because this detection methodology relies on a database of known signatures, it is unable to detect new malware or the signatures of malware which has not been added to its database.

3.1.1 Polymorphic Malware

Polymorphic viruses disguise themselves during the duplication process to avoid signature-based detection. Typically, these types of viruses are embedded into other executables and when an infected executable is launched, the virus located other executable and embeds itself into the newly found executable [7].

Polymorphic viruses, encrypt the body of the virus using different encryption keys to both obfuscate the code, and to avoid binary fingerprinting. This helps avoid detection through signature recognition and disassembly; as the code's signature is changed through the use of a new encryption key, and disassembly is prevented through general encryption of the code. A metamorphic element is maybe included with modern polymorphic viruses to assist with evasion of signature detection through the use of a mutation engine, which generates different decryption algorithm for new version of the virus.

This type of malware has the unique requirement of needing to pack and unpack itself. This has led some AV vendors applying a blanket blacklist to known packers which are commonly used by malware writers. While this is not a true detection method and may result in false positives, mythologies have been developed to detect polymorphic by executing part of the suspected application in a sandbox, and analysis the output of instruction writing to memory, if these writes generate an application, various static methodologies can be applied to the output for a more reliable detection methods [7].

3.1.2 Metamorphic Engines

Metamorphic engines can use a number of techniques to disguises malware; however, they do not change its functionality. The various known techniques used by metamorphic engines are:

- **Garbage insertion**: A common form of obfuscation, code is inserted that will not execute or executes without effecting the outcome of the code. By injecting this "garbage code" the application hopes to change itself enough not to be recognised against existing signatures.
- **Formal Gramma Mutation**: Formal Gramma mutations follows a set of rules represented by *G* to create syntactically correct code by generating all possible permutation of a tokenised string, however system call analysis can still be used to detect this type of malware [8].
- **Instruction Substitution**: Instructions may be "substituted" with equivalent instructions or groups of instructions.
- **Registry Swap**: A weak for of metamorphism, where registry are changed. By observing the opcode and using wildcards to against the registries, signatures may still be used to identify this type of metamorphic malware.
- **Subroutine Permutation**: malware my compile the subroutines in difference sequences, to create a new binary signature.
- **Transposition**: Where instruction do not have dependencies, it is possible to reorder their execution.

It should be noted that simply recompiling the code using a different compile, or enabling certain compile time optimisations, the compiler may perform transposition, instruction substitution, or remove dead code, thus with very little work an attacker may achieve some metamorphic properties in their next compiled malware. This has a particular bearing in APTs where an attacker develops a specialist piece of malware for a single target. The same code could be compiled a few years later, using a newer compiler which could potentially create a different signature.

3.2 Malware Detection and Classification

A particle interesting work by Lin and Stamp [9], poses a different type of challenge for malware detection, where it was accepted that metamorphic viruses could be detected using hidden Markov models (HMM), yet they set about creating a metamorphic engine which could evade detection. They concluded that "spectral analysis", where the spectrum consisted of instruction sets of infected code, was ineffective at detecting malware metamorphic malware. Despite this Pechaz et al., was able to achieve 90% accuracy in detection and classification [10] using Markov models.

While there are maybe challenges associated with static binary analysis, program binaries offer little meaningful identification to identify their intention. Application metadata where set may provide some clues to the legitimacy of an application but its intent is usually completely obfuscated from the user. The paper "Functions-based CFG Embedding for Malware Homology Analysis", identified that in most cases the source code for malware involved in APTs are never revealed, thus compiler identification is the first step toward disassembly, and ultimately enables various techniques to be used to analyse the code. However, even once disassembled, this is only the beginning of the challenge in identifying the malware's intent, how to defend against it, or its origins. Comments, class names, function names and variable names are often removed during compilation as part of the compiler's optimisation. Using various tools, it is often possible to dissemble a binary and view system calls and calls to external assemblies to ascertain some information on the assembly. From this information, it is here that it is possible to use various graphing methods to create an identity for a binary, although this is not without its challenges. Muzzamil Noor [1], concluded that static malware analysis could not fully unlock the intent of a malware application, due to obfuscated and encrypted binaries.

The further compound this, the common practices of library reuse, which is encouraged under the SOLID Principles and OOP has increased the binary simi- larities of applications. Furthermore, the rise of open-source projects has accelerated this trend in modern software allowing developers to not only reuse libraries, but to borrow individual functions from open-source software and include it in their own applications. Therefore, both malware families and legitimate application are more likely than ever to share libraries [11] and snippets of code.

A combination of graph functions can be used to check for similarities between known malware, and new binaries. Alrabaeem proposed introducing a new code representation named the semantic integrated graph, build upon three existing graph representations control flow graphs (CFG), registry flow graphs RFG, and program dependency graphs. As noted, accuracy of these predictions suffers where the devel- oper is not constrained by OOP principles, where larger functions are used or func- tions with different semantics are placed in the space source file. It is reasonable to summarise code similarly based on library reuse is not effective when code modularity is low.

Further complication with similarity indexing, occurs with binary analysis due the use of obfuscation, and complier optimisation. Compilers may choose to remove dead code, rename variables, replace instruction with equivalent instruction, while obfuscator can do this as well as rename registers, rename variables, move methods, and insert new methods that yield no effect.

3.3 Homology Analysis

Homology Analysis of malware is important especially in where APTs are concerned as it helps to identify, not only the similarities between malware such as which family it may belong to, but also provides hints to the origins of said malware. The paper "Functions-based CFG Embedding for Malware Homology Analysis", the

Table 3 mCrab block-level attributes [12]

Type	Attribute name
Block-level attribute	String constants
	Numeric constants
	No. of transfer instructions
	No. of calls
	No. of instructions
	No. of arithmetic instructions

use of Deep Learning Convolutional Neural Networks (CNN) and improved short-term memory networks (LSTM) for malware homology. The team behind the paper, proposed a new system named mCrab, that made use common CFG and ACGF (Attributed Control Flow Graph) analysis to generate training data for the CNN which could later be used for classification, and APT grouping. It is worth noting that the mCrab prototype build upon Xiaojun et al. [12] improved methodology which confirmed that inter-locking block attributes were unnecessary and block-level attributes of ACFG would suffice (Table 3).

It is interesting that mCrab is able to use such a simplified ACFG vertices to encode as feature for training of the neural network, yet return such good results.

The creators of the mCrab prototype acknowledge code snippets and tricks may be used by different malware writers, therefore even if there is a high similarity between code the code malware may belong to a different APT group. Citing their experiment WannaCry ransomware, which their model pointed towards the Lazrus group, which security experts all agreed was sceptical. While the model performed well in malware classification, the reuse of libraries and code by different APT groups, and possibly mimic of APT groups make meant the model was not suitable for identifying the malware's author.

CNN models have good identification capabilities with sparse visual learning datasets. Work by Zhang et al. [13], demonstrated that malware can represented to a visual learning model, by converting opcode sequences to an image to identify and classify malware. The OpenCV library erode the image and then dilate them with the effect of denoising the images while making them both conspicuous and retaining useful information. This particular technique is interesting as the opcode profiling was able to work by selecting 2-tuple opcode sequences only. Thus, reducing the amount of data required for the CNN model to process, and greatly reducing the processing power required.

The MSCS (Malware Classification using SimHash and CNN) algorithm presented by Ni et al. [14], selected a methodology to present SimHashes as grayscale images to train a CNN model for malware identification. SimHashing was developed by Charikar [15], and hashes similar values to similar hashes, and allows us to use the hammering distances between hashes to determine similarity as a numeric value. As with other CNN models, the binaries are disassembled, encoded and then used to train the CNN model for malware classification. However, due to the performance of SimHashing the entire opcode sequence can be used for the encoding. To fix the issue of binary length, each malware code is encoded using SimHashs of equal length. From this each SimHash can be converted to an 8-bit grayscale pixel and arranged on a dot matrix to resemble something similar to a QR Code.

All research yields the same issues when tackling obfuscated code. Advanced techniques are required to disassemble the binary, for feature encoding. However, this does not mean that mean that binaries that are obfuscated cannot be classified by DNNs, as other feature can be picked out, at a small cost of reduce confidence in the model's prediction.

3.4 Packers

Packers are not exclusively used for malware. In principle, packers are a form of protection for a section of code. A packer simply encrypts, compresses, or both in a binary file. When the file is executed the packers, stub will unencrypt/decompress the packed section, and load the executable into memory, in its unpacked format to run as normal.

While most software developers have no need for packers, some software developer may choose to use packers to minimise the size of the binary file. While it can be argued that a developer may use a packer to protect their intellectual property, this is seldom done in practice, as there are a number of code obfuscation programs freely available. However more advanced packers are able to proxy the Import Address Table, making it difficult to use memory dumps to reverse engineer the code. Advanced packers will use code virtualisation, and anti-sandboxing techniques to hinder the detection of malware [16].

Essentially packers make it difficult to perform static analysis on malware, resulting in packers being target instead as a means to determine the safety of a file. Where actual understanding is required of malware manual analysis if often required.

3.5 APTs

ATPs pose a specific problem as identified by Paul Giura and Wei Wang in A Context-Based Detection Framework for Advanced Persistent Threats [17], which hypothesised that while traditional Threat Modelling and Threat Trees are helpful, they cannot discover, and thus defend against all eventualities. In the case of APTs they identify that a combination of persistence, patients and along with complex kill chains, means that attacks can go undetected. Therefore, they proposed the Attack Pyramid Model, but this model does not indicate if there is one induvial acting alone or if the attacker is a large state funded agency. Rivera and Hare [18] summarises that larger organisation and capable individual actors can directly engage at any level, making it much harder to determine the attackers' capabilities in cyberspace than it is in the real world, however acknowledge powerful enteritis such as governments, and the military are likely to wield more powerful cyber-capabilities even if it is against less traditional targets, such as private individuals.

"Feature analysis for data-driven APT-related malware discrimination" argues that "*features that characterize the malware used in APTs might help security analysts*" [19]. Where, Haddadpajouh [20], also confirms that there is a significate overlap in threat "*actor's, tactics, techniques*, and procedures.

3.6   Visual Learning

As many deep learning approaches to malware identification using CNN, are based on BytePlot visual learning, where either the binary's features are encoded as pixels in a greyscale image [21], or the entire assembly is encoded as an image. There are several other benefits with visual learning, namely continued advancements in the field, good performance with sparse datasets, and well-known approaches to transfer learning. However specific benefits to visual learning to malware detection and categorisations are noted in the paper "Robust Intelligent Malware Detection Using Deep Learning" [22], which states "*image processing techniques require neither disassembly nor code execution, it is faster in comparison to the Static and Dynamic analysis*". This feature is particularly useful for IoT devices which may have limited resources as disassembly can be complex task, and memory intensive. This technique allows a binary to be converted into an image and the resulting image can be analysed remotely, or even on the same device. Beyond this as confirmed by Vinayakumar et al. [22], this methodology, as the benefit of being cross-operating system compatible. Both the encoding of the binary can occur on any operating system provided that the same algorithm is used, and the resulting image may be analysed on any operating system, provided the model has been trained appropriately. This universal approach to malware analysis has a great potential for further development and refinement, without operating system specific lock-in.

Various models such as ResNet-50 [23], have confirm the validity of this approach for malware classification. ResNet-50 obtained a 98.62% accuracy rate, on a dataset consisting of 9339 sample of malware from 25 different families. Furthermore, where large datasets are available, experiments have shown excellent results with 98% accuracy for identification of malware using popular obfuscation techniques [22].

The scope of visual learning is increasing, and disassembly, to reveal further information such as CFGs, RFG, and FCG (function control graphs), along with other features may be encoded into images so that a visual learning model can classify a binary.

3.6.1 Feature Encoding for Visual Learning

It is accepted that there are generally three types of data which can be encoded:

- Categorical: Things that can be sorted into categories.
- Quantitative: Things that can be counted.
- Ordered: Things that can be sorted into a sequence.

Normally feature encoding can occur in multiple dimensions depending on the number of features. However, in visual learning feature encoding in often planar, and in the case of both BytePlot and LocalEntropy these render well on a 2D Surface, enabling common Computer Vision (CV) libraries to be used with in visual learning.

While the most popular uses of computer vision are:

- Object detection: Where is an object in an image
- Object classification: Categorisation of objects in an image
- Object identification: What object is in an image.

These popular usages of CV, can be used to detect the presence of malware, the malware's family, and even the exact type of malware, using existing technologies provided that the binary features have been encoded in a way that the model can learn.

3.7 Transfer Learning

While this paper does not explore the use of Transfer Learning. It recognises the importance of Transfer Learning in the context of APTs, where data on emerging threats or perpetrators may be sparse. Or where the model may become outdated quickly due variance in the scenarios (environments) that the model is applied to; such as variances in Operating Systems versions, or new functions being created common libraries/SDK which malware can call in different ways. As recognised by Yang et al. [24] a major challenge in the use of Machine Learning is that models often do not work well in new task domains. To improve any Deep Learning Model, transfer learning can be used to improve the baseline performance, and reduce model-development time. This is of particular interest where the source domain and target domain differ, or data cannot be shared between domains.

In the context of APTs for target domain improvements, Multiparty Learning lends its self well to a scenario where "*several organizations*," possibly government agencies "*wish to build a model together, while keeping their sensitive data private*" [24]. As stated in the article Learning Privately from Multipart Data [25], Hamm rightly asserts that two agencies may want to collaborate on malware detection model without revealing or sharing their user data. However, their requirement for privacy could mean that they are unable to share quality training data and thus are unable to benefit from each other's knowledge. Thus, being unable to train a learning model.

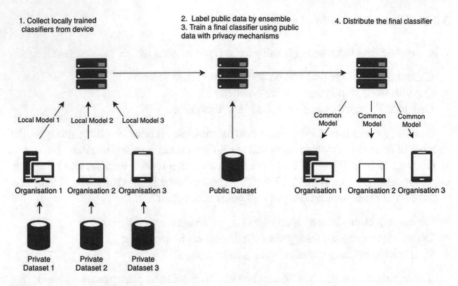

Fig. 2 Workflow of multiparty learning (adapted from Hamm et al. [25])

Having well-designed AI Systems without the needed training data is like having a sports car without any energy. [25]

Hamm summarises that discrete pretrained learning-models can be shared, and that combined with publicly available data a new learning-model can be created and then distributed without sharing any private data (Fig. 2).

In this workflow, it was proposed that local models are used to generate labels for the public dataset which in turn is used to train the final model. While this should provide the privacy required to protect each organisation's dataset, the paper acknowledges privacy within Transfer Learning *"comes at a loss of classifier performance"*. Thus, it is imperative that the performance of any local model is optimised before being used to generate the labels for the public dataset.

As previously noted, the model for identify and classifying malware may greatly benefit from Heterogeneous Transfer Learning, where meta data, behaviour and other attributes may greatly increase the baseline performance of the model by providing supplemental information thus increase the model's confidence. The relatedness of missing application vendor signing or an invalid signing and malware, maybe used to increase prediction confidence. However, a significant problem with this approach is finding related latent factors, and often manual translation between domains. In addition, a real world the target domain is likely to have both a larger feature space, and varying probability distribution, thus reducing the effectiveness of models created in test environments.

4 Experiment

As seen above, without an extensive prelabelled datasets, naming either the source or author of a particular malware sample, it is not possible to train existing commercially available models to predict the author of new malware. Furthermore, the number limited number of samples, with number of shared features means that training a model becomes especially difficult without the use of complex techniques such as Deep M-sparse encoders, or Bi-Level Sparse Clustering.

As such the experiment shall focus on the identification and classification of malware using existing Deep Learning Models, specifically in the field of Computer Vision as this has yielded good results, in various applications such as mCarb.

4.1 Method

To develop a Deep Learning model capable of identifying malware using a supervised learning; a diverse dataset containing different malware samples is required, ideally many of the malware samples will be from the same family.

Furthermore, a training dataset, validation dataset and test dataset are required. The training dataset is used to train the data model, while a validation dataset is used for fine tuning. Finally, the model is tested with unseen dataset data to determine the model's accuracy.

The malware dataset was gathered from Malware Bazaar [26], which holds a collection of Malware Samples. While a control dataset, normal files (binaries free of malware) were gathered from reputable vendors, which was legally licenced and installed on a test machine, running antivirus. The control data set consisted as binary samples from an installation of Microsoft Windows 10, Microsoft Office, Adobe Creative Cloud and Blackmagic Designs switching software. While this is not a particularly diverse sample of software it should provide the DNN with enough data to recognise normal files.

The Training dataset consisted of 81 Dynamic Link Libraries (DLLs) infected with malware, 130 executables infected with malware, 21 normal DLLs and 21 normal executables.

4.2 Experiment Configuration

There are various approaches to working malware analysis, however a key consideration with any approach is the ability to keep the test environment free from adverse effects of the malware being analysed.

A virtualised test environment, was selected using VMWare WorkStation using an Intel based Apple Mac as the host system. The selected guest operating system was Ubuntu 20.04 LTS, this decision was made as the sample malware are PE and DLLs which are designed for use on the Microsoft Windows Operating System. File System, and clip-board sharing was disabled, and the hosts network adapter was disabled. The guest used a USB-Bridge to connect to a dedicated ethernet adapter that has access to a dedicated internet connection.

For simplification, all samples were categorised as either Normal, or Malware. It was felt that file category of benign should be ignored as software in this category may or may not be malicious, and as such would only serve to create additional noise in an already sparse dataset. Grayware was also removed from the dataset, due to controversy over its definition. Both adware and spyware fall into the category of grayware [5], which many would argue to be malware.

The malware samples were taken from MalwareBazaar a free online database of malware samples.

Disclaimer: For simplicity only Windows base executables were used in the dataset, to reducing the training time, to make the dataset more manageable, and reduce risk of adverse effects on the host systems. Due to anti-bot restriction, gathering samples automatically was prohibited.

The files were examined in "Detect It Easy", and the following were obtained from the file's makeup:

- Compiler
- Packer
- File signature.

This was done to ensure that the files were indeed DLLs and Windows executables, and had not been mislabelled in the database.

The samples were encoded in their entirety as a PNGs using a BinVis.io. BinVis.io offers multiple options for encoding files. The files were encoded using cluster, for both ByteClass and Entropy. Due to the nature of the generic visual learning classes to be used cluster was chosen as the encoding method over scan. Traditionally scan encoding would have been more appropriate for word, or tabular encoding which would be suitable to other types of deep learning models.

Three experiments were executed against two different automated ML systems; Apple's Create ML, and Google Cloud's AutoML.

4.2.1   Experiment 1

The training data was labelled, with a single label; either malware, or normal depending on its content. Training completed for both Create ML and AutoML. With Auto ML showing good precision and accuracy at the 90% confidence threshold.

4.2.2 Experiment 2

The training data was labelled, normal where no malware was contained. The rest of the data was given the known malware family names, depending on the family of malware that was contained in the binary. Where the family name was unknown, the data label was entered as "malware".

Only, AutoML was able to train the model successfully, but due to the lack of samples for each family of malware, the results were unsatisfactory, *see Appendix B* Table 7.

4.2.3 Experiment 3

In an attempt to denoise the image, and provide a pattern that the neural network could learn. Adobe Photoshop 2021 was used to apply a Gaussian Blur of 1.2 pixels and the image was then convert to grayscale, similar to previous BytePlot methodologies [21]. Unfortunately, this the automated ML servers failed to complete. No models were trained as a result.

5 Critical Discussion

Preliminary results, show that given the correct algorithm it is possible to detect the potential presence of malware using ByteClass Image Analysis. ByteClass Image Analysis performed better at the 90% confidence threshold than Entropy Image Analysis. Whilst this is not conclusive, and additional research is required to validate these finding, through the use of a larger more varied dataset it provides the grounds for continued research into Clustered ByteClass Image Analysis for malware detection (Table 4).

This was contrary to predictions, in which the author believed that Entropy Image Analysis would have provided better results in malware, due to the common practice of encrypting or packing malware with in the binaries. The ByteClass Image Analysis performed better across the confidence range, with a higher precision (Figs. 3 and 4).

One of the biggest issues with deep machine learning is the difficulty in understanding what it is that the machine has learned. Essentially the encoding of a model is

Table 4 Results from Googles AutoML image object classification for malware detection

	Provider	Image type	Confidence threshold	Recall (%)	Precision (%)
Malware detection	Google AutoML	ByteClass	0.90	84	95.5
		Entropy	0.90	68	89.5

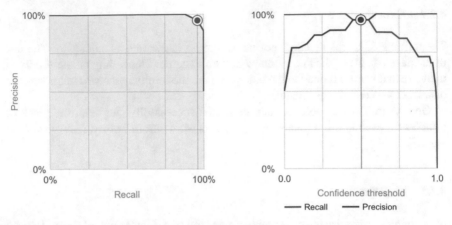

Fig. 3 ByteClass precision for detecting malware in Google AutoML

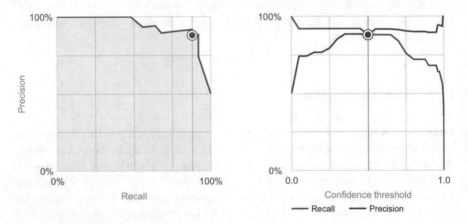

Fig. 4 Entropy precision for detecting malware in Google AutoML

one-way. On multi-feature models where reduction techniques are used, the resultant model may ignore parameters that humans would find to be valuable information.

The most simplistic description for the use of DNNs is based on the principle of defining a learning outcome, for a particular set of inputs and then fine the neural networks model's parameters until that outcome is achieved. However, in practice this is usually a lot more complex. A DNN developer would carefully consider the input data, and encode them as features appropriate for the type of model they intended to use. The developer may then choose to mix models in a hierarchy, and define the outputs, before fine tuning the model. This requires not only an understanding of Neural Networks, but an understating of the various models available, and the mathematics behind them. Furthermore, with traditional machine learning, some experience is required with one or more software development languages, and knowledge of common libraries is essential. With the advancement of ML, a new breed of automatic ML is enabling those with limited knowledge of AI and ML or even software development to create effective models for various application. As these tools are still in their infancy, they require vast quantities of data for training. In the short term this will drive a quest for more and more malware samples, which may not be forthcoming where malware has been developed by APT groups for specific targets.

One could theorise that an APT group develops a unique piece of ransomware to target a sewage treatment plant. As this is unique to this particular sewage plant, there would not be a known file signature, and thus there is little need to pack the malware, thus avoiding detection through entropy analysis. Using common practices in commercial software development such as the use compiler optimisations, and off-the-shelf obfuscators, can slow down the reverse engineering of malware enough for an attacker to achieve their intent, as often it is cheaper to pay the ransom the financial penalties of distribution to business.

5.1   Malware Acquisition

With current automated ML technologies for image analysis heavily focusing on vision learning for object classification and object detection, many providers are defaulting to CNNs for image analysis. These early for automated ML implementations require a large sample set for each classification to train a model. Acquiring these datasets has become more viable for enterprises but less viable for individuals. Free online malware databases, are implementing anti-bot systems to defend themselves from DDoS attacks and scrapping. Which makes gathering sample from free services in an automated way extraordinarily time consuming, and impractical for gathering more than a few hundred malware samples. One should note that free malware databases tend not to be very large, and as such may not have a large sample of malware from the same family or category. What's more many malware databases have not been updated in over a decade, while other malware database websites have been abandoned or moved without redirection.

To make matter worse, malware databases does not rank on the first five pages of the world's most popular search engine Google, and only small malware databases list on the world's second most popular search engine Bing.

There isn't a standard for implemented a malware catalogue or database, and as such this makes searching for malware samples across providers difficult, as binary comparison or disassembly is required to ensure duplicates are entered to a researcher's dataset. This issue was also noted by O'Shaughnessy [27], with AV Vendors descriptions of malware, stating that standards do exist for providing malware descriptions in the form of Malware Attribute Enumeration and Characterization (MAEC™), which is now on version 5.0.

These factors are clearly a hinderance to the development of malware detection and categorisation technologies and techniques outside of commercial and academia. But work by Zhang et al. [28], and Paranthaman and Thuraisingham [29], outline common feature extraction that could be used malware cataloguing and indexing. Namely byte-file N-Gram, number of imported APIs (dlls), and mnemonic instruction counts. These are all common values used for feature encoding for malware classification through deep learning, and thus could contribute to a standard for creating a malware database which is indexable via common properties. Taking this a step further, and indexing CFG as NoSQL documents would not only help researcher, without disassembly skills to collaborate with malware experts, it would effectively reduce the learning curve thus attracting more engineer to the field of malware analysis, increasing the man power available.

5.2   Static Malware Analysis

There are so many ways in which to analysis binary code, and ultimately allow for binary code fingerprinting that the choice can be become overwhelming. Although not strictly addressed in the paper, it is important to note that binary code fingerprinting, is important for defending against malware and that many malware detection and removal application rely on binary code fingerprints to detect and remove malware.

There are various frameworks, tools and practices for binary code analysis, many of which can compare binaries for similarities, decompile binaries into OpCode, generate flow control graphs, view memory tables and much more. These frameworks, tools, and practices allow humans to reverse engineer a binary and fingerprint it. While it is not important to list all of the methods for fingerprinting binaries, it is important to know that it is possible to fingerprint both binaries and functions. It is also important to understand that various features of a binary can be converted to values upon which mathematical formulars can be executed to find similarities, or combined and hashed to create unique identifiers. These values are known as features and there are common features for all binaries, and can be encoded into many different formats for different types of neural networks.

5.3   Zero-Code ML

Stacked convolutional neural networks, are resource intensive and dependant on hardware capabilities. New multi-core GPU, enhancement to the $\times$ 64 architecture and SoC technology has allowed for further development in this field, allowing for a larger number of fully-connected convolution layers as well as a larger final softmax layer. As hardware continues improves, we are able to create deeper CNN, which intern are able to learn increasingly complex features. However, with deep neural networks we can run into issues with overfitting. Popular overfitting techniques include L1/L2 regularization, dropout and data augmentation [30]. In the case of automated ML, the decision on which cost function to apply to the neural network, will be made for us as will the decision regarding the function used to forward propagate the signal to reduce the occurrence of vanishing gradients. However there no guarantee that a particular type of neural network will be used by an automated ML solution. As such a less than ideal, trail-and-error approach to data augmentation is required to optimise results. Typically, with CV ML, skewing, blurring, shifting, rotating or applying noise may be used to increase the size of the training. However, the case of generated images, where pixel position carries meaning, moving a pixel does not seem prudent. This leaves the introduction of blue and noise into the image as possible types of data augmentation.

In the case of the automated ML experiments, the introduction of even a small amount of Gaussian blur cause training of the model to fail, demonstrating that although, this technique was used to enhance results for BytePlot encoding of particular features in the mCarb experiment. This is likely to do with the model generated by the automated ML and should a different model be generated it is possible that a small amount of Gaussian blur be useful enhancing the results. Therefore while, automated ML serves as a good starting point for validation of a theory, ultimately it is not refined enough to be used in real world scenarios, where the model may require manual finetuning. As it stands the only method available to fine tune an automated ML model would be to provide more data, or employ transfer learning where allows. In the case of this experiment, it became exceedingly clear that there simply were not enough images to feed the neural network, and thus malware classification was not possible.

5.3.1 Feature Encoding

The main input to automated ML is labels and features. In the case of computer vision the features are learned from the image, using label and boundary box attributes. As this is one of the few inputs available for image object classification and image object detection it is essential that those images are of a high quality, and the label/boundary box attributes are correct.

One of the appeals of Visual learning and specifically CNNs is their ability to perform feature extraction; it is able to break large images into smaller one and extract distinct features. While humans are fantastic at object detection, object recognition, and pattern recognition, there we have our limitations and CNNs are able to find district features and patterns where we would miss them. Thus, encoding vast quantities of data into an image and hoping for the best is a distinct possibility, and this can be seen in this experiment as we encode the entire binary into an image, even though we are aware that only sections of the binary would need to be encoded. It is certain that a more targeted approach to encoding the features would result in good accuracy as demonstrated in "*RMD: Malware Variant Detection Using Opcode Image Recognition*" [28] where features were carefully selected and the encoded into the image.

Speed

Early attempts to encode features into for computer vision, were done in black and white, however the amount of data that can be encoded into a black and white image is limited, as such modern encoding is completed in colour. This enables a higher bit count per pixel, and thus data rate thanks to SIMD extensions found on modern CPUs.

In the context of encoding binaries into images, data need not be lost regardless of the colour depth, but due to library optimisation the encoding usually happens at 8 or 12-bit, however there is nothing stopping us from using 16-bits per channel. In the case of the experiment the images generated were PNGs with 8-bits per channel, providing a good contrast between pixels. Not only did this mean that images could be encoded quickly, they could also be decoded quickly on commodity hardware, and thus the CNN could be trained quickly. By taking advantage of SIMD functions on the Intel base CPU it was possible to encode the image locally, in under one second each.

Continuous Fracture Space-Filling Curve

When encoding binaries into a planar visual representation, it is necessary to select an algorithm that would provide good clustering of features while maintaining byte order, as such scan encoding was immediately discarded, and as with other malware researchers using image object classification the experiment used the Hilbert Curve to encode the final image. Research suggest that the Hilbert Curve performed better than Gray-Coded and Z Curves and producing in worse case and average scenario [31].

Due to the standard libraries being used by the automated ML applications, it was essential to provide an image which would be as optimal as possible for image object detection and classification. The paper *"Image-based Malware Classification: A Space Filling Curve Approach"*, build upon existing work by S. Mukkamala demonstrates that the Hilbert Curve works for BytePlot image when the researcher had control over the DNN model being used.

5.3.2 Transfer Learning

Researchers in researching deep learning for malware classification, have tended to shy away from transfer learning, and instead seem content with training new models from scratch. This is a concern, especially in the domain of image classification for CNNs, where there are well-known approaches to transfer learning:

- Pertained network as a classifier
- Pertained network as a feature extractor
- Fine-tuning.

These approaches are designed to save time, and improve the model's precision, and with so many researchers encoding features into images using similar algorithms i.e. Hilbert-Curve, for BytePlots, Entropy and ByteClass encoding, and even building their research upon existing research, it would be reasonable to use an existing model to save time on model training, as this is one of the most resource and time intensive task in machine learning.

With so many researchers creating models and sharing their findings online, it would make sense for them to also share their models. Particularly as the infrastructure is already inplace to do so.

In a game of cat and mouse, malware authors are clearly sharing tooling, code snippets, and classes, to stay ahead.

Lifelong Machine Learning

A question remains around the validity of Lifelong ML in the context of malware detection and classification. While it is true that large amounts of data become available over time, technology moves at such a pace that a piece of malware developed in the 1980s bears little resemblance to malware developer today. At a binary level 16-bit ($\times 3086$) code has a narrow address space, fewer registries and a smaller instruction set than a modern 64-bit ($\times 64$) code, and for the most part is not even compatible with modern operating system.

Would there be value to a DNN learning malware samples from the 1980s for classification? It could be argued, that after a period of time, the data and thus sample set becomes obsolete. However, that it not entirely the point behind Lifelong ML; Lifelong ML, seeks to learn from different task in multiple domain [, p. 197]. By learning from different tasks, among several domains Lifelong ML aims constantly update the knowledge base with transferable knowledge. In the case of malware classification, combine various knowledge domains, to gain increased our prediction accuracy. For instance, we can use domain knowledge CFGs (Control Flow Graphs), RFGs (Registry Flow Graphs), and FCGs (Function Call Graphs) to complement each other to build a new model, that can accurately classify malware. By combining these knowledge domains, we are able to create skip-layers which can reduce the time the model takes to perform a prediction, based on a high signal value in earlier layers.24

As this data and knowledge can build up overtime the model stands a good chance of recognising new malware, even if it is unable to classify it, due it belonging to a new family.

5.4 APTs

It is obvious that due to the nature of APTs there is a great interest in detecting and defending against them. However much of the recent research is design to detect APTs and APT malware at the network level [32].

With treat actors, actively monitoring the success of each other's campaigns and adopting, successful techniques, the similarity between concurrent campaigns is increasing [20]. Ultimately Haddadpajouh, et al. sums up the problem with one sentence:

Digital traces and cyber evidence are much more fragile, and actors are well educated in this domain. [20]

This essentially means that APTs are sufficiently educated to disguise their malware, deliberately injecting code which carries known signatures of other APTs. Their technical ability should not be underestimated, as they will go through great length in both the cyberspace and the real world to hide their true identity.

Beyond this APTs organisation often use both legal systems and geopolitics to their advantage, masking banking transaction in countries with more secretive banking regulations, or routing traffic through countries on less favourable term with the victim.

Looking at the landscape of Cyberthreat attribution, we quickly see that this is a difficult task with very research papers available on the subject. Part of the reason for this could be that, as Cybersecurity advances, and employs multi-level defences, APT organisations are becoming more specialised; Working together as a team to form kill-chains, where one organisation my specialise in the writing of ransomware, and another APT organisation specialises in the delivery of the ransomware to the victim. In this scenario, we now have tactics and practices of two distinct organisations, along with learnt tactics, and behaviours of other APT, and code which with lineage to multiple APT organisations. Finding discriminating features between APT organisations under these circumstances, would be extremely challenging.

However, threat attribution for malware is not impossible, and Haddadpajouh et al. [20], was able to use Fuzzy logic against a multiple-view of malware dataset, with a 95% accuracy. While this model used a small number of APT groups and a relatively large sample set for such a small number of categories (the 5 APT groups), it demonstrates the principle is viable and that APT attribution while difficult is possible.

6   Conclusion

It is clear that the discovery of new Malware is currently too slow to protect organisation against APTs, APTs are currently able to keep ahead of the organisations in the development of successful malware. Attackers are creating more dynamic malware as exhibited by Silver Sparrow [33] where the payload for the malware appeared to be missing, presumably to be dynamically generated or retrieved at a later date, based on some sort of trigger.

Deep Learning has been applied by various organisation at the network parameter, and network interfaces (endpoint protection) with some success. However, as there are many entry points to a system; media drives (USB-Drives, CD Drives, etc.), Bluetooth connections and so forth, which are not secured by endpoint protection, it is necessary to perform binary analysis to help safe guard the system against malicious applications, and files.

It is conclusive that Deep Neural Networks are suitable for Malware detection and classification. A recent bias towards Computer Vision learning systems, appears faster and less resource intensive than traditional Static Analysis Detection Techniques. However, while common implementation of disassembly and graphing is a complex task and unlikely to be performed on end devices, it is likely that further SoC (System on a Chip) implementations will only continue to add the required power to turn this into a trivial task. Deeming the factor of speed and resources mute in the future.

While Deep Belief Networks could be used with unlabelled data, providing a possible edge over conventual neural networks for identification and classification of Malware [34], identify the source of APTs still presents a sizeable challenge. Without a larger number of co-operating agencies gathering and sharing data on APTs, it would not be possible to generate enough data to perform the supervised learning required for the Dense Neural Network to learn the patterns to confidently identify the source of an APT from Static Analysis alone. Thus, making creating a paradoxical problem, where while it is possible for ML to confidently identify the source of APT. It is not possible unless enough training data can be provided (Political issues). As such DNNs may not be suitable for identifying the creator/author of a particular piece of malware, due to three main reason:

1. The lack of data on malware authors, thus the model does not have the enough datapoints or labels to attribute a piece of malware to an author.
2. CV machine learning for style identification is firmly rooted in the art industry, which is unsuitable for current binary feature encoding methods.
3. Common tools which are used for compiling, obfuscating, and encrypting a binary, coupled with the use of shared libraries and code snippets, can mask the style of the author or collection of authors. Beyond this, software developers are increasing following standardised patterns and using code linters, helping them to conform to a standardise coding style.

Due to rapid adaptation of Malware creators, it is important that Deep Learning models are kept up to date, therefore any implementation of a neural network will require transfer learning to ensure knowledge between domains it transferable, and that new neural networks can be trained quickly in response to additional information becoming available.

Currently zero code implementation of Zero Code vision systems is not suitable for malware identification. These commercially available models are designed for object recognition, with clear boundaries.

7 Future Work

While there is no sparsity of malware samples, assigning malware to a specific author or group of authors is difficult, and requires human assessment. As such, an interesting concept which could possibly be borrowed from Computer Vision would be style recognition, and kinship understanding [35]. Expanding on BytePlot techniques it would not be incomprehensible that Computer Vision may be able to detect and identify coding styles, even after obfuscation and compiler optimisation. As noted by [35], style is usually attributed to high-level abstract concepts, however much work in this area focuses on nuances for feature encoding. These nuanced encoding of malware features has enabled automated categorisation of malware, and indeed this has been achieved with good accuracy as previously stated. Shared code snippets, libraries and functions, as well as the utilisation common, obfuscators, packers, and compilers has enabled datamining techniques such as random forests of give a 94.2% accuracy in identification for malware classification [36], where there are large sample groups. However, it should be noted that that depending on the library used much less satisfactory results can be seen using random forest, and this same study show that neural networks, performed poorly with less than a 27% accuracy rate, thus the selection of Neural Network Library, can have a huge impact on the accuracy of the model.

The current research into malware identification and classification does not lend itself well to style classification. Style classification usually focuses of discriminative feature representation for classification, however where data is sparse; namely the lack of prelabelled malware identifying the author, introduces noise. Thus, making it difficult for neural networks to differentiate differences and greatly increasing the challenge of fine tuning.

References

1. Muzzamil Noor HA (2013) Anticipating dormant functionality in malware: a semantics based approach. In: 2013 International symposium on biometrics and security technologies. IEEE, Chengdu, China, pp 20–23
2. Chen P, Desmet L, Huygens C (n.d.) A study on advanced persistent threats
3. Collins (2018) Collins English dictionary, 13th ed. Collins, Glasgow
4. Corfield G (2020) Newcastle University, neighbouring Northumbria hit by ransomware attacks. [Online] https://www.theregister.com/2020/09/08/newcastle_northumbria_universit ies_cyber_attack/. Accessed 08 Sept 2020
5. Hampo JA (2019) Comparative analysis of malware detection techniques. LAP LAMBERT Academic Publishing, Beau Bassin, Mauritius

6. Sonnemaker T (2021) 'Liar's dividend': The more we learn about deepfakes, the more dangerous they become. https://www.businessinsider.com/deepfakes-liars-dividend-explained-future-misinformation-social-media-fake-news-2021-4?r=US&IR=T, cited on 20th September 21

7. Kasina A, Suthar A, Kumar R (2010) Detection of polymorphic viruses in windows executables. In: Contemporary computing, 95 (IC3 2010. Communications in Computer and Information Science), pp 120–130

8. Zbitskiy PV (2009) Code mutation techniques by means of formal grammars and automatons. J Comput Virol 5:199–207

9. Lin D, Stamp M (2011) Hunting for undetectable metamorphic viruses. J Comput Virol 7:201–214

10. Pechaz B, Jahan MV, Jalali M (2015) Malware detection using hidden Markov model based on Markov blanket feature selection method. In: 2015 International congress on technology, communication and knowledge (ICTCK), pp 558–563

11. Alrabaee S, Debbabi M, Shirani P, Wang L, Youssef A, Rahimian A, Nouh L, Mouheb D, Huang H, Hanna A (2020) Binary code fingerprinting for cybersecurity.Springer Nature, Switzerland

12. Xu X, Liu C, Feng Q, Yin H, Song L, Song D (2017) Neural network-based graph embedding for cross-platform binary code similarity detection. ACM Digital Library, Dallas, TX, USA

13. Zhang J et al (2016) IRMD: malware variant detection using opcode image recognition. In: 2016 IEEE 22nd international conference on parallel and distributed systems (ICPADS). IEEE, Wuhan, pp 1175–1180

14. Ni S, Qian Q, Zhang R (2018) Malware identification using visualization images and deep learning. Comput Secur 77:871–885

15. Charikar M (2002) Similarity estimation techniques from rounding algorithms. Princeton University, Princeton

16. Roccia T (2017) Malware packers use tricks to avoid analysis, detection. [Online] https://www.mcafee.com/blogs/enterprise/malware-packers-use-tricks-avoid-analysis-detection/. Accessed 12 March 2021

17. Pual Giura WW (2012) A context-based detection framework for advanced persistent threats. In: 2012 international conference on cyber security. IEEE, Alexandria, VA, pp 69–74

18. Rivera J, Hare F (2014) The deployment of attribution agnostic cyberdefense constructs and internally based cyberthreat countermeasures. In: 2014 6th international conference on cyber conflict (CyCon 2014), vol 6, pp 99–116

19. Liras LFM, Soto ARd, Prada MA (2021) Feature analysis for data-driven APT-related malware discrimination. Comput Secur 104(May 2021):1–15

20. Haddadpajouh H, Azmoodeh A, Dehghantanha A, Parizi RM (2020) MVFCC: a multi-view fuzzy consensus clustering model for malware threat attribution. IEEE Access 8:139188–139198

21. Conti G, Dean E, Sinda M, Sangster B (2008) Visual reverse engineering of binary and data files. In: Visualization for computer security, 5210 (VizSec 2008. Lecture Notes in Computer Science), pp 1–17

22. Vinayakumar R et al (2019) Robust intelligent malware detection using deep learning. IEEE Access 7:46717–46738

23. Rezende E et al (2017) Malicious software classification using transfer learning of ResNet-50 deep neural network. In: 2017 16th IEEE international conference on machine learning and applications (ICMLA). IEEE, Cancun, Mexico, pp 1011–1014

24. Yang Q, Zhang Y, Dai W, Pan SJ (2020) Privacy-preserving transfer learning. In: Transfer learning. Cambridge University Press, Cambridge, p 215

25. Hamm J, Cao Y, Belkin M (2016) Learning privately from multiparty data. In: Proceedings of the 33rd international conference on machine learning, pp 555–563

26. MalwareBazaar Database (2021) MalwareBazaar Database. [Online] https://bazaar.abuse.ch/browse/. Accessed 15 March 2021

27. O'Shaughnessy S (2019) Image-based malware classification: a space filling curve approach. In: 2019 IEEE symposium on visualization for cyber security (VizSec), pp 1–10

28. Zhang Y et al (2017) Based on multi-features and clustering ensemble method for automatic malware categorization. In: 2017 IEEE Trustcom/BigDataSE/ICES, Sydney, NSW, Australia, pp 73–82
29. Paranthaman R, Thuraisingham B (2017) Malware collection and analysis. IEEE, San Diego, CA, USA
30. Krohn J, Beyleveld G, Bassens A (2020) Chapter 9 Improved deep networks. In: Deep learning illustrated, a visual, interactive guide to artificial intelligence. Pearson Addison-Wesley, s.l., pp 131–155
31. Moon B, Jagadish H, Faloutsos C, Saltz JH, January F (2001) Analysis of the clustering properties of the Hilbert space-filling curve. IEEE Trans Knowl Data Eng 13(1):124–141
32. Zhao G, Xu K, Xu L, Wu B (2015) Detecting APT malware infections based on malicious DNS and traffic analysis. IEEE Access 3:1132–1142
33. Smith A (2021) Mysterious malware discovered on 30,000 new Macs—and researchers have no idea what it was designed to do. [Online] https://www.independent.co.uk/life-style/gadgets-and-tech/malware-new-macs-m1-b1805582.html. Accessed 22 Feb 2021
34. Ding Y, Chen S, Xu J (2016) Application of deep belief networks for opcode based malware detection. IEEE, Vancouver, British Columbia, Canada, pp 3901–3908
35. Jiang S, Shao M, Xiong C, Fu Y (2019) Style recognition and kinship understanding. In: Wang Z, Fu Y, Huang TS (eds) Deep learning through sparse and low ranking models. Elsevier, London, pp 213–249
36. Mahajan G, Saini B, Anand S (2019) Malware classification using machine learning algorithms and tools. In: 2019 Second international conference on advanced computational and communication paradigms (ICACCP). IEEE, Gangtok, India, pp 1–8

Information Security Accountability in the Cloud Computing Context—A Comprehensive Review

Zahir Ahmed Al-Rashdi, Martin Dick, Rahma Ahmed Al-Rashdi, and Younis Al-Husaini

Abstract Accountability is a main concern for information security within cloud computing; it represents the trust in service relationships between clients and cloud service providers. Without evidence of accountability, a lack of trust and confidence in cloud computing is to be expected from decision-makers. Furthermore, a lack of accountability is considered as an added level of risk, especially since a client's essential services are controlled and managed by a third party. Therefore, this new outsourcing paradigm increases the challenge of maintaining data security and confidentiality, supporting data and service availability, and demonstrating compliance. This chapter presents a literature review on IS accountability. It sets out the different definitions of IS responsibility from the existing literature. This chapter reviews information security and cloud issues, and related security issues and how they relate to IS accountability in the context of cloud computing. The concept of Cloud computing along with the different types of cloud services is thoroughly described. The factors of accountability are also reviewed in this chapter. This chapter is made up of two main sections. The first section will focus on information security and explore IS Accountability in cloud services provisions. The second section will elaborate on the conceptual drivers of IS Accountability. These factors are Transparency, Responsibility, Assurance and Remediation. Finally, accountability for Cloud computing service relationships and all other aspects associated with IS Accountability will also be identified and explained.

Keywords Cloud computing · Information security · Information security accountability · Cloud service provision · Outsourcing · Accountability elements

Z. A. Al-Rashdi (✉)
Information Security Department, Sultan Qaboos University, Muscat, Oman
e-mail: zaher21@squ.edu.om

M. Dick · Y. Al-Husaini
RMIT University, Melbourne, VIC, Australia

R. A. Al-Rashdi
Keio University, Tokyo, Japan

© The Author(s), under exclusive license to Springer Nature Switzerland AG 2021
R. Montasari et al. (eds.), *Challenges in the IoT and Smart Environments*,
Advanced Sciences and Technologies for Security Applications,
https://doi.org/10.1007/978-3-030-87166-6_8

1 Introduction

This review provides an explanation of what Information Security Accountability (IS Accountability) in a Cloud computing context is, and how government organisations can ensure that it is present in Cloud computing service relationships. Although information security and privacy in relation to Cloud computing has received a great deal of attention from researchers in the field of information systems (Infosyss) [14, 71] and information security [72], yet, information security accountability in Cloud computing has not been studied in great depth. Furthermore, many of the studies concentrate on technical aspects such as encryption and preventive controls. Although technical aspects for cloud security and privacy have been actively researched, the focus on detective controls in relation to cloud accountability and auditability is scarce. Encryption and other privacy protection techniques will only manage a part of this problem. In addition, there is the problem of ensuring that security obligations are implemented by cloud service providers. According to Gartner, globally end-user expenditure on public cloud services would rise 18.4% to $304.9 billion in 2021, up from $257.5 billion in 2020 [23]. Thus, the enormous growth in moving businesses to Cloud computing, due to its flexibility, cost-effectiveness, scalability, and the perceived benefits of transference of data security and the absence of a specific Cloud computing accountability framework, highlights the growing need for research in this area. Research is needed into accountability and auditability of cloud service providers to affect both preventive and detective measures in ways that promote transparency, governance, and the accountability of the cloud service providers.

1.1 Evolution of Accountability

The principle of accountability is found in the well-known OECD Guidelines; in the laws of the European Union ("EU"), the EU member states, Canada and the United States; in emerging governance such as the APEC Privacy Framework and the Spanish Data Protection Agency's Joint Proposal for an International Privacy Standard. The Organisation for Economic Co-operation and Development (OECD) established accountability as a principle of data protection in 1980 and since then has played an increasingly important and visible role in privacy governance. The emergence of the accountability principle places responsibility on organisations as data controllers to comply with measures that give effect to all of the OECD principles [75]. In the European Union, the principle of accountability initially considered privacy protection including the implementation of processes by organisations, which in turn assessed how much data was to be collected, the usefulness and the usability of the collected data and the protection level required to ensure information security. The transfer of data outside the EU has been managed to ensure safe transfers of sensitive and personal data, which was addressed in the EU accountability principle in the data governance section [77]. In February 2009, The Spanish Data Protection Agency's established a basis for data transfers and created the Joint Proposal for an International Privacy Standard, which included the principle of accountability [78, 79]. The office of the Privacy Commission of Canada established the first principle of accountability in 2009 under Canada's Personal Information Protection and Electronic Documents Act (PIPEDA) that soon became part of the law that relates to processing, storing and transferring data domestically and outside the Canadian border [76]. In the United States, the government has acted initiatively to enhance the principle of data protection and accountability by imposing legal obligations, Under the Gramm-Leach-Bliley Act, and the Safeguards Rule, enforced by the Federal Trade Commission, requires financial institutions to have a security plan to protect the confidentiality and integrity of personal consumer information [89]. The Center for Information Policy Leadership (CIPL) is working hard on the accountability approach in the digital world because much data has been migrated to cloud environments. CIPL has paid more attention to improving accountability in the public cloud and mobile services by presenting the main risk element associated with these two cloud environments [29, 62].

1.2 What is Information Security Accountability?

Accountability is a core concern for information security in Cloud computing, it represents the trust in service relationships between clients and cloud service providers (CSPs) [9]. Without evidence of accountability, a lack of trust and confidence in Cloud computing [11] is developed by decision-makers and considered as an added level of risk, which means a lack of accountability increases [9], especially since a client's essential services are controlled and managed by a third-party. Consequently, this new method of outsourcing renders the process of maintaining data security and privacy, supporting data and service availability, and demonstrating compliance far less transparent [27]. This makes it difficult for users to understand, influence and determine what security obligations are implemented by CSPs.

Many researchers indicate that accountability should be given more attention and treated as a high priority issue in terms of security [45, 50] as it affects the quality of service (QoS) [5, 56] as well as the grade of service (GoS) [5]. Generally, most users are seeking Assurance that their QoS and GoS requirements are satisfied and that their operations are not hindered due to congested cloud resources. Providing the required assurance measures and guarantees for both QoS and GoS is a challenging task. Furthermore, accountability—along with trust—are two major concepts that are considered foundational for potential users wishing to embrace cloud services.

In the remainder of this chapter, Sect. 2 outlines the importance of information security for organisations and investments into information and communication technologies. We also discuss cloud computing and related issues such as: cloud deployment models and cloud services, adoption drivers, and current issues of cloud computing. Then, Sect. 3 provides insight into information security accountability to discuss in detail the relationship between accountability and cloud computing within an organisational context. Section 4 covers information security accountability conceptual factors. Lastly, the conclusion is shared in Sect. 5.

2   Literature Review

2.1   Importance of Information Security

Information is considered the most valuable and crucial asset to any organisation, and hence, must be properly protected [25]. The use of information is evolving at an unprecedented rate and Escherich [21], a principal research analyst at Gartner, emphasises that information within an organisation is vitally important and has to be thoroughly protected. However, downloads and the use of many different types of software that process this information, brings along more threats to organisations [38]. For example, attackers can use malicious tools to gain access to various valuable information resources and services such as identities and credentials. This information can be used by attackers to gain profit illegally [22]. It is also evident that the need for information security for both personal and institutional use has rapidly increased due to the proliferation of communication media, electronic storage and transmission of information [93]. Some important reasons for this growth are due to the "increase in electronic applications in businesses as well as in daily life, the sharing of information on network systems, the accessibility of information from many points, the increasing threat of loss of information, and most importantly, the increases in personal and corporate losses" [20].

2.2   Organisations and Investments in Information Technology

Many researchers have proposed that all investments of business operations and IT should be integrated into their business values and should be aligned with organisational strategy [16, 41]. For example, Croteau and Raymond [17] confirm that the investment in IT and business processes should be coordinated to achieve a proper strategic plan with a good integration process. Ju et al. [32] propose that the correlation between strategic factors, organisational factors, and technology alliance have a great impact on an organisation's competitiveness. However, Markus [41] argues that approaching success and improvements in terms of functions and performance is difficult whether or not IT is required. There is a resistance to investing in technology in most organisations due to a lack of information, human resources, and cost of implementing internal security management systems.

2.3 Organisations and Information Security Concerns

Information systems security remains the most challengeable task to IT leaders, executives and professionals [19]. Maintaining Infosys security in organisations is more than just a technical matter. There are other aspects of Infosys, such as organisational "grounded principles and values" [19], which need to be considered. There are various studies in the literature which emphasise that Infosys security is more effective in terms of management if it goes beyond technical aspects [57]. For example, Puhakainen and Siponen [57] state that employee refuses to comply with information security policy should be considered as a real information security threat. Furthermore, Straub and Welke [69] state that there are several values to be measured in terms of protecting information resources at any organisation. Segev et al. [64] outline that the main key of protecting Infosys security is not technical but that it should be accomplished by studying the key managerial elements featuring each organisation.

Tan and Hunter [74] suggest that a mix of social and organisational factors must be considered as effective values to be employed by Infosys stakeholders. These organisational and social factors indicate people's assumptions, accountability, and values towards Infosys security issues [47]. Keeney and Keeney [34] state that re-evaluating these social factors can help to discover some hidden objectives. Trompeter and Eloff [82] argue that in addition to technical and organisational factors, ethics and personal accountability must be considered throughout the implementation of any ISMS. However, In the context of this review, the researcher will focus mainly on the challenges of information security accountability in the cloud computing context.

2.4 Cloud Computing Service Provisions

Cloud computing relates to the use of online computing services and is considered an on-demand IT service or product based on the business model. Users and businesses can use software and hardware through cloud services, including SaaS, PaaS and IaaS with the management of third parties at a remote location [68]. Characteristics include: manageability, access method, performance, multi-tenancy, scalability, data availability, control, storage efficiency [83], advanced security technologies [67], on request allocation and reallocation of resources, virtualised storage and networking facility, enabling sharable resources "as a service" model, the flexibility of moving an organisation's data through data centres, cost-effectiveness, reducing the responsibility of maintaining data locally, and resources made customisable on the web [92]. In addition, the computing resources and data are automatically maintained through software that is managed and controlled by CSPs [63].

According to the National Institute of Standards and Technology (NIST), Cloud computing is "a model for enabling ubiquitous, convenient, on-demand network access to a shared pool of configurable computing resources (e.g. networks, servers, storage, applications and services), that can be rapidly provisioned and released with minimal management effort or service provider interaction"[42].

2.5 Cloud Computing Deployment Models and Cloud Computing Services

The cloud infrastructure can be subdivided into four layers: the physical layer, the infrastructure layer, the platform layer and the application layer. In addition, cloud computing is made up of four models: public, private, hybrid and community. Each of these models is divided into three service models: Software as a Service (SaaS), Platform as a Service (PaaS), and Infrastructure as a Service (IaaS) as demonstrated in Figs. 1 and 2 respectively [4, 12].

The rapid growth of Cloud computing has created a paradigm shift in technology since the whole IT infrastructure has become available as a service including within smart city sectors [3]. Despite the fears of losing control by different users about data stewardship especially health and financial data, there are a number of notable commercial and individual cloud computing services, including Google (Email Service), Microsoft Azure and Yahoo [84]. Ko et al. [36] state that Google, Microsoft (Azure) and Amazon (EC2/S3) are the current prominent cloud providers in the world. For example, Microsoft Office 365 provided by Microsoft is the most popular case of SaaS service provided to the public, Google Apps is a good example of PaaS and Amazon Web Services is a good example of an IaaS [46].

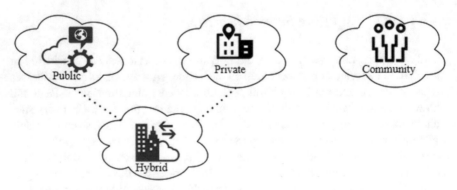

Fig. 1 Cloud computing deployment models

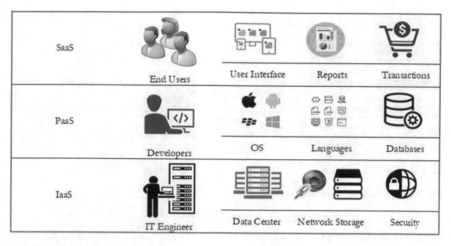

Fig. 2 Cloud computing services

2.6   A Motivation for Cloud Migration

Cloud computing relates to the use of online computing services and is considered an on-demand IT service or product based on the business model. Users and businesses can use software and hardware through cloud services, including SaaS, PaaS, and IaaS with the management of third parties at a remote location [84]. Characteristics include: manageability, access method, performance, multi-tenancy, scalability, data availability, control, storage efficiency [35, 88], advanced security technologies [35], on request allocation and reallocation of resources, virtualised storage and networking facility, enabling sharable resources "as a-service" model, the flexibility of moving an organisation's data through data centres, cost-effectiveness, reducing the responsibility of maintaining data locally, and resources made customisable on the web. In addition, the computing resources and data are automatically maintained through software that is managed and controlled by the CSP [63]. Overall the sharing of resources represents the main benefits of Cloud computing by sharing large pools of resources such as compute cycles, or virtual CPUs (VCPUs), storage and software services [44, 59]. However, sharing resources increases concerns towards security with end users, particularly with respect to data or applications hosted in the cloud provider's data centres [33].

2.7   Current Issues for Cloud Computing

In recent years, the demand for migration to clouds is ever-increasing due to the growing number of personal data including bookmarks, photographs, media and music files, are accessed remotely via a network [94]. Cloud computing has expanded

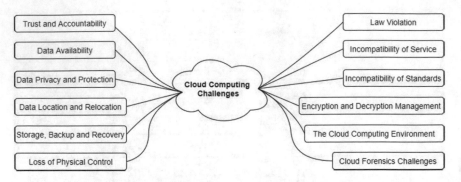

Fig. 3 Cloud computing challenges

into one of the fastest-growing portions of the IT industry and has become a promising business concept where a huge amount of information—for both individuals and enterprises—is placed. This transformation of data distribution and storage in the cloud has generated a real concern towards data privacy and data protection. It has also raised questions about how safe the cloud environment is. This question is considered by most organisations before deciding on deploying their business into the cloud [70].

After reviewing more than eighteen case studies by the researcher about cloud computing, between 2009 and 2021, it was revealed that there are many challenges and security issues associated with cloud migration [58, 92]. These challenges, as shown in Fig. 3 and discussed in subsequent sections, have a significant impact on users' decisions to move their activities to the clouds.

2.8 Cloud Computing Challenges

From the above-mentioned cloud computing challenges presented in Fig. 3, it is clear that migrating data to the cloud is not an easy task to achieve in terms of decision making and implementation. Decision-makers should ensure that accountability mechanisms are in place to provide clients with control and transparency over data in the cloud. This includes enabling the customer to choose and select its cloud provider based on the criteria of reliability and IS responsibility and vice versa. When this happens, trust and accountability are fully implemented. In addition, data availability must also be ensured where client data are usually stored in 'blocks', usually in different locations and on different servers. This would contribute negatively to the availability of data and could constitute a genuine concern for the availability of uninterrupted and uninterrupted data provisions. Confidentiality and data protection are other aspects that must be taken into account, as confidentiality and data protection include confidentiality, integrity and availability. A proper practice, privacy policies and information system procedures must be clearly stated in the service level agreement (SLA) [60] to assure the client of data safety. The integrity mechanisms should be implemented by the CSP to ensure that data will not be modified to any extent. The client should be made aware of any data loss or change [13] and the right information should be made available to the right people [66]. Furthermore, the CSPs should inform the client about the data location (where the data will be stored, for example, in Australia or the US or India), this should be clearly mentioned in the SLA [18, 40]. Furthermore, the customer must be aware of the possibility of moving or transferring their data from one cloud to another [18]. The consumer needs to be sure about the whole cloud computing environment. The physical and political environment surrounding the data centre and how safe it is (data location) The consumer needs assurance about the availability of adequate data storage systems in place, at least RAID (Redundant Array of Independent Discs) systems. This can be accomplished by having the latest technology in storage, backup and retrieval systems. Recovery systems should be regularly updated and maintained to restore information as early as possible in the event of a malfunction. This is part of the emergency plan agreed to in SLA to manage events and respond to incidents.

Consumers are generally concerned about the loss of physical control as data and resources are shared with CSP in whole or in part. Sometimes governments never comply with privacy laws when it comes to accessing other people's data, indicating a clear breach of the law [37]. This includes the limitation of collecting citizens' data that resides/goes into storage, the duration of time to keep data and some financial, banking protocol, and that the customer's data should remain within the country. From a technical perspective, consumers are also concerned with the issue of compatibility, such as the incompatibility of services between different service providers. This is especially true when the customer decides to switch between cloud providers. For example, Microsoft, Cloud and Google Cloud are both incompatible with one another and incompatible with current standards and code of practice across the various clouds. Another conflict is emerging between the CSP and its associated consumers in terms of encryption and decryption management. It's about who should take control of encryption and decryption, the customer or the cloud provider? The consumer has to be clearly informed about the process and procedures the CSP will follow in case of data breach and how the investigation or what is known as a cloud forensics process will be maintained. This is a real challenge, as more than one-third party will host the data at any time and everyone will share the responsibility of hosting customer data. Cloud forensic investigations require complex procedures and special tools, along with the exceptional skills a digital forensic investigator must possess to conduct this type of investigation [1, 6]. Since digital forensics is one of the key elements of law enforcement (LEA) [7], CSP collaboration is vital. Therefore, the relationship between digital forensic investigators and CSPs must be framed to enhance trust between the two parties and ensure the highest level of preservation of the integrity of digital evidence [8]. In addition to the above-mentioned points, the studies also revealed some logistical issues such as service level agreement issues (SLA) [43], costing models, charging models [48], what to migrate [15], and cloud interoperability issues.

As part of this review, this chapter will focus on the challenges of IT security accountability. Information security practitioners believe that cloud ownership is the first "building block" that each organization must implement to improve cloud data protection. Ko et al. [36] and Lynn et al. [39] believe that cloud accountability is one of the evolving issues in cloud security and needs to receive the greatest attention. In addition, scholars also believe that accountability is incorporated directly with all the other information security challenges outlined above, and has a great impact on the implemented mechanisms, which in turn ensures responsible decision-making towards information security management and protection of data [24, 36].

3 Information Security Accountability

The review revealed four main components of Information Security accountability in relation to Cloud computing service provision, and showed that in order to be an accountable organisation, the four components should be fully implemented. As can be seen in the descriptions of these factors, as illustrated in Fig. 4, there are inherent interactions between the four components (Responsibility, Assurance, Transparency and Remediation) which means that they should all be addressed simultaneously; Uniqueness implementing each component is likely to cause failure in relation to IS Accountability. ISPs who wish to be held accountable should be aware of these four

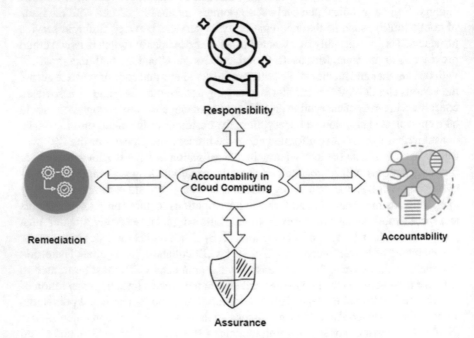

Fig. 4 The four central components of Accountability in the existing literature

core elements of accountability and be prepared to demonstrate to customers that they have achieved them. However, this does not mean that there is a singular method for implementing all four factors or any individual factor—the nature of the organisation, its industry context, the type of data collected, the business model, and the potential risks that data usage raises for clients all have an impact on the method chosen for implementing these factors [24, 52]. The following section details four different areas associated with accountability. First, the organization and accountability of end-users are briefly presented. Second, there is a discussion on what accountability means. Lastly, the conceptual factors for IS responsibility are listed and discussed.

3.1 Organisations and End Users' Accountability

People will remain the most responsible party in mitigating human risks posed by malicious tools [2]. There is social responsibility in combating threats and therefore, all parties should work together to mitigate the risks of malicious threats and that includes governments, ISPs, end-users, and international bodies.

3.2 What Is Accountability and How Does It Relate to Cloud Computing and ISMSs

Accountability is a global term that has been used and presented for a number of years in computer science, finance and public governance, and is becoming more incorporated into business regulatory programs. However, recently the term accountability has emerged in terms of a worldwide privacy and data protection framework [55]. The research effort on accountability has produced a considerable number of definitions. These definitions embody different spheres of accountability research. Both academics and practitioners have different views and interpretations of accountability. For example, accountability in computer science according to scholars is referred to a limited and imprecise requirement that is met by reporting and auditing mechanisms [52]. Yao et al. [91], considers accountability as a way of making the system accountable and trustworthy by the combination of mechanisms. Accountability should be approached by having mechanisms that concentrate on a broader scope of the legal issues [89]. For example, Vedder and Naudts [85] suggested calling for accountability mechanisms that transcend the mechanisms that are inherent in regulation. Authors like Jaatun et al. [28] proposed to develop accountability mechanisms that would diversify processes, non-technical mechanisms and tools that would support accountability practices. Rush [61] defines accountability as the reporting and auditing of mechanisms and obligating an organisation to be answerable for its actions.

Vithanwattana et al. [86] identified accountability as the process of tracking users' activity in a continuous manner while accessing resources in the system. The users would be tracked against the systems they accessed, which systems they were granted access to, what type of information they accessed, the amount of data transferred during their access, time spent on the system and when they disconnected from the system.

Muppala et al. [44] refer to accountability as the adherence to accepting the ownership and responsibility towards all actions in a standardised way as regulated by an acknowledged organisation such as the Organisation for Economic Cooperation and Development (OECD) who published privacy guidelines in 1980. However, Ko et al. [36] consider accountability as one out of the four components of trust in Cloud computing. The remaining three are security mechanisms (e.g. encryption), privacy (the protection of personal or confidential data not to be exposed) and auditability. A reasonable definition for accountability has been provided by the Galway project of privacy regulators and privacy professionals. To them, accountability is defined to as the commitment towards safeguarding personal information with an obligation to act as a responsible steward by taking responsibility for protecting, managing and appropriating use of that information beyond mere legal requirements, and to be held accountable for any misuse of that information [76].

The Centre for Information Policy Leadership in the United States has identified accountability in relation to privacy as "the acceptance of responsibility for personal information protection. An accountable organisation must have in place appropriate policies and procedures that promote good practices which, taken as a whole, constitute a privacy management program. The outcome is a demonstrable capacity to comply, at a minimum, with applicable privacy laws. Done properly, it should promote trust and confidence on the part of consumers, and thereby enhance competitive and reputational advantages for organisations" [78, 79].

Pearson [53] stated certain conditions need to be met in order to provide an improved basis of trustworthiness in the Cloud computing environment, which automatically enhances accountability.

This review uses the definitions of accountability offered by several authors [76, 86]. These definitions cover the central components of accountability namely Transparency, Responsibility, Assurance, and Remediation. For example, organisations are required to demonstrate a certain level of acknowledgement and assumption of Responsibility by introducing or having in place appropriate policies and procedures along with promoting good practices for correction and remediation in case of failure and misconduct [52]. In addition, in terms of data protection, organisations should be held responsible for any decision made about the protection of data by report, explanation, enhancing Transparency, considering liability and be made answerable for the consequences [44]. Governance and ethical dimensions along with promoting the implementation of practical mechanisms are all parts of accountability, whereas legal aspects and guidance should be interpreted for data protection [77]. The above definitions are used as the basis of this comprehensive review because the definitions incorporate the protection of organisational assets and personal privacy.

However, none of these studies has mentioned the role of the ISMS in controlling all aspects of accountability in Cloud computing. This indicates that the need to employ an ISMS in order to control all aspects of security challenges in Cloud computing is a trend in today's world. Accountability in relation to cloud service provision needs to be examined and investigated with regards to its integration with ISMSs.

Therefore, accountability and ISMSs are interconnected with each other, since both of them are designed to satisfy the same approach towards information security management and data protection. Their main goals are to ensure the protection of organisational assets and personal privacy regardless of the difference in processes and procedures followed to achieve each approach.

4 Conceptual Factors

The Centre for Information Policy Leadership has identified accountability as "a demonstrable acknowledgement and assumption of responsibility for having in place appropriate policies and procedures, and promotion of good practices that include correction and remediation for failures and misconduct. It is a concept that has governance and ethical dimensions. It envisages an infrastructure that fosters responsible decision-making, engenders answerability, enhances transparency and considers a liability. It encompasses expectations that organisations will report, explain and be answerable for the consequences of decisions about the protection of data. Accountability promotes the implementation of practical mechanisms whereby legal requirements and guidance are translated into effective protection of data" [77].

As mentioned previously, this review uses the definition of Accountability offered by several authors [76, 78, 79]. In particular, these definitions cover the central components of accountability: Transparency, Responsibility, Assurance and Remediation. These definitions are used as the basis of this review because they incorporate the protection of organisational assets and personal privacy. Figure 4 shows the overall interactions between these four key components in terms of achieving accountability. It should be noted that the double arrows indicate that the four factors interact with each other and are not necessarily independent of each other.

4.1 Responsibility

Responsibility in the context of accountability for Cloud computing service provision is the acknowledgement and assumption of responsibility by CSPs that they have introduced or have in place appropriate policies and procedures [26]. Responsibility is achievable by ensuring the existence of obligatory and enforceable written data privacy policies and procedures that reflect applicable laws, regulations and industry standards. The accountable CSP should develop, implement and communicate to clients a set of data privacy policies that are informed by appropriate external criteria recognised by laws, regulations or the industry's best practices [81]. In addition, the accountable CSP should be prepared to provide clients with [87] and also design and deploy a set of procedures to implement effective and practical written policies according to the circumstances of each organisation—such as what data is collected, how it is used, and how systems and organisations are connected.

Responsibility is considered one of the most important factors of IS Accountability to adequately manage the relationship between CSPs and clients. The accountable organisation (CSPs) should have a data privacy program in place to establish, demonstrate and test its accountability. Each organisation should demonstrate a level of Responsibility and a willingness to be accountable for any misconduct in its data practices, policies and procedures, which should be implemented based on external legislative criteria [30], and generally accepted principles or the industry's best practices. All policies and procedures must be approved at the highest level of the organisation, and senior management should demonstrate their commitment towards motivating Responsibility, which in turn encourages accountability.

4.2 Assurance

Assurance, in terms of Accountability for Cloud computing service provision, is to comply with governance and ethical measurements along with promoting the implementation of practical mechanisms that are commonly considered key parts of accountable processes and procedures [54]. In addition, Assurance is considered as the main tool of evidence that provides valuable information to risk management where this evidence would be used by providing confidence to stakeholders that the qualities of service and stewardship with which they are concerned are being managed and maintained appropriately [54]. The following factors are part of Assurance in relation to accountability for the provision of Cloud computing services:

- Staffing and delegation
- Education and awareness
- Mechanisms to manage IS accountability in the cloud computing environment
- Ongoing risk assessment and mitigation
- Program risk assessment oversight and validation
- Event management and complaint handling
- Internal enforcement.

Appropriately trained personnel will ensure the validity of the CSP's privacy program and assign the right resources to the right personnel [77]. Small and medium-sized organisations should ensure that these delegations are in line with their specific activities and circumstances, such as the nature, size and sensitivity of their data holdings. Once properly implemented, the client-CSP relationship will be enhanced;

CSPs will provide their employees with appropriate training and be assigned responsibility for the privacy program [30]. Education and awareness is another element of assurance that improves IS accountability to the CSP.

An effective education and awareness program will ensure that staff of an organization and on-site contractors are kept informed of data protection obligations. Such ongoing education and awareness will enhance the CSP's employees' capabilities and increase their understanding of the essentiality of protecting clients' data to avoid data leakage consequences such as job dismissal. It is clear that this process will increase the level of trust between clients and their CSPs. The implementation of a number of CSP mechanisms improves the assurance factor, which increases IS accountability. Implementing mechanisms directly impacts on managing IS accountability in the cloud environment. The review identified several mechanisms that, according to the researchers, support IS empowerment in a cloud computing environment [30, 31]. The mechanisms are considered tools and activities to implement and monitor IS accountability objectives. In addition, Jena and Mohanty [31] have discussed the importance of cloud auditors being used in online dispute resolutions (ODR) in the cloud environment where data is remotely controlled as part of the compliance process during the Remediation stage. The importance of cloud auditors as an effective mechanism was discussed by Jaatun et al. [30]. For example, mechanisms to clarify compliance with respect to extraterritorial legislative requirements and provide a list of certifications required should be handled by the cloud auditor, as they are considered to be the main actor to perform audits and certifications, to monitor accountability levels of cloud providers and to make sure that collection of implicitly collected data is made transparent [30].

Another aspect of assurance is the constant assessment and mitigation of risks. CSP and their associated clients agreed that to be an accountable CSP, processes are needed to understand the related risks to privacy that may occur from the implementation of new solutions, products, services, technologies or business models. The results of ongoing risk assessment and mitigation should be taken into account in the measures taken by the organization to mitigate potential client risks. In addition, these organizations should further demonstrate how these decisions are taken and what actions are taken to mitigate the risk. By having such steps in place, including precise processes and procedures to arrest ongoing risk and mitigation, CSPs will be perceived positively by clients in terms of accountability, and are likely to be branded as a trustworthy party. In addition, having a solid program risk assessment oversight and validation in place enhances the trust and confidence of CSP accountability as it assures clients that constant reviews are set in place. Ongoing reviews of an organization's confidentiality and accountability program need to be considered by clients and CSPs in order to be viewed as a responsible organization. This will ensure that the needs of an organization are consistently met through sound data management and protection decisions that promote and respect privacy outcomes. Information security practitioners believe that a review of the exchange of such programs between PSPs and clients will improve trust and validate CSP programs, which in turn will lead to a responsible partnership. In addition, security experts believe that implementing complaint management and handling systems is seen as an added value to cloud clients. Decision-makers believe that a responsible organisation should put in place procedures that effectively respond to requests for information, complaints and data protection breaches [30]. A timely response to any inquiry, complaint or violation in terms of data protection will establish an image of support between clients and CSPs, which in turn will strengthen accountability between them. The matter of support between clients and their CSPs can sometimes become complicated and often leads to contract termination, which means a loss of trust between clients and CSPs. The in-house application is another assurance factor that enhances IS accountability in cloud computing environments. IT experts believe that accountable organisations should have in place methods to enforce internal policy, ensuring that any breaches to those internal data protection rules by employees—such as IS Accountability practitioners—property or misuse of data, are subject to sanctions, including discharge [9, 77]. In this case, this internal application is directly linked to CSPs and the reinforcement of these aspects of the application will increase the likelihood that customers choose to use this CSP.

As a whole, assurance and accountability are interrelated. Based on corporate culture, each organization must establish performance systems to be viewed as a responsible organization, and the following characteristics represent successful performance systems: (1) they are consistent with the organisation's culture and are integrated into business processes; (2) they assess risk across the entire data life cycle; (3) they include training, decision-making tools and monitoring; (4) they apply to outside vendors and other third parties, to ensure that personal data obligations are met no matter where the data is processed; (5) they allocate resources in places where the risk to individuals is greatest; and (6) they are a function of an organisation's policies and commitment. In Europe, North America and Asia–Pacific seal programs are used where they play the role of third-party accountability agents, which provides external oversight by making Assurance and verification reviews a requirement for participating organisations.

4.3 Transparency

To ensure transparency in the context of accountability for providing cloud computing services, a series of issues need to be addressed. The results of each review, including changes in rules and procedures, should be communicated to customers in a clear and timely manner [73]. Information should be properly communicated to client organizations and regulators in a rigorous and cost-effective manner [51]. As part of this process, the outcome of assessment measures or audits should be reported to the appropriate employee within a client organisation, and where necessary corrective action should be taken [27, 49]. Transparency involves reporting and explaining decisions taken to protect data. It also means that acceptance of liability and remedies are clearly presented to customers [10]. Transparency between clients and their CSPs is an essential element towards achieving accountability in Cloud computing, as most clients want to know who is handling their data as well as how, where and when it is used. Sharing such information between clients and CSPs increases trust, which increases accountability [30]. A level of transparency must be demonstrated among clients and organizations. Clients should have rights to the data collected by editing it. They also have the right to stop using certain data where it is not appropriate or to correct the data collected where it is inaccurate. However, there may be limits on disclosing information in certain circumstances.

4.4 Remediation

To implement corrective actions that ensure accountability for cloud service delivery, there is a need for a range of corrective action processes. According to The Centre for Information Policy Leadership [84] remediation is "the method by which an organization provides recourse to persons whose privacy has been put in jeopardy." In this context, a responsible organization should use its best practices in remedial action and redress in the event of failure and misconduct [52, 65]. In addition, an accountable organisation should have a specific remediation mechanism that suits each organisation according to their data holdings, and the way the data is used and appropriated for a specific issue. These mechanisms should be readily accessible to clients, and the lead organization should be able to handle complaints in an effective and efficient manner. The redress mechanisms would vary depending on the culture and the industry. Therefore, decisions about redress should be made locally.

However, these remediation mechanisms would need to be developed in consultation with a range of experts, regulators, civil society, and representatives of both public and private sector organisations [90]. Corrective actions complement accounting processes and procedures to ensure business continuity in the event of a malfunction. When failure occurs, individuals should have access to a recourse mechanism. For instance, a third-party agency might be needed to address and resolve the failure that has occurred. Customers must be aware of the processes and procedures that must be followed in the event of failure.

5 Conclusion

This review has sought to understand how IS Accountability in Cloud computing can be conceptualised. Initially, this review used an extensive analysis of the literature relating to Cloud computing and accountability for information security to develop a model of the key conceptual elements (Responsibility, Transparency, Assurance and Remediation) (see Sect. 2.15) relating to this issue. The objective of this review was to understand what an organization needs to do to achieve IS responsibility in a cloud context. It should be noted that this differs from information security because an organization considered to be responsible for information security may still have corresponding violations. In fact, some aspects of information security and accountability, such as remedial actions, may never come into play if such a breach does not occur. In order to determine if an organization is responsible for its information security, the first step is to identify and define the core elements of IS accountability. The review found that these four elements were viewed by IS accountability practitioners as core elements of IS accountability.

This reviewer provided a more detailed definition of IS accountability based on the revealed literature. This definition will assist in the research of IS Accountability by providing a common understanding of the concept. The researcher examined the meaning of IS Accountability to determine whether the four components of the earlier IS Accountability in Cloud model could be expanded. Achieving IS Accountability is a complex task for any organisation. The first step is to understand the elements of IS Accountability. It is important to realise that though there are four elements, the level to which any element needs to be implemented in an organisation, is highly context-dependent. A level of transparency for an organization with an acceptable level of accountability may not be sufficient for another organization. Overall, it needs to be understood, that the four-element model is not prescriptive and that it must be used in a context-sensitive way that is dependent on the needs of the specific organisation that is attempting to achieve IS Accountability.

References

1. AL-Husaini Y, Al-Khateeb H, Warren M, Pan L (2018) A model to facilitate collaborative digital forensic investigations for law enforcement: the royal Oman Police as a case study. In: Paper presented at the 2018 cyber forensic and security international conference, Nuku'alofa, Kingdom of Tonga, pp 21–23
2. Abraham S, Chengalur-Smith I (2010) An overview of social engineering malware: trends, tactics, and implications. Technol Soc 32(3):183–196
3. Ahmadi-Assalemi G, Al-Khateeb H, Epiphaniou G, Maple C (2020) Cyber resilience and incident response in smart cities: a systematic literature review. Smart Cities 3(3):894–927. https://doi.org/10.3390/smartcities3030046
4. Ahmed ZE, Saeed RA, Mukherjee A (2019) Challenges and opportunities in vehicular cloud computing. In: Cloud security: concepts, methodologies, tools, and applications. IGI Global, pp 2168–2185
5. Akintoye SB, Bagula A (2019) Improving quality-of-service in cloud/fog computing through efficient resource allocation. Sensors 19(6):1267
6. Al-Husaini Y, Al-Khateeb H, Warren M, Pan L, Epiphaniou G (2020) Collaborative digital forensic investigations model for law enforcement: Oman as a case study. In: Security and organization within IoT and smart cities. CRC Press, pp 157–180
7. Al-Husaini Y, Warren M, Pan L (2018) Cloud forensics relationship between the law enforcement and cloud service providers. In: Paper presented at the CWAR 2018: proceedings of the 17th Australian cyber warfare conference
8. Al-Husaini Y, Warren M, Pan L, Gharibi MA (2019) Cloud forensics investigations relationship: a model and instrument
9. Al-Rashdi Z, Dick M, Storey I (2017) Core elements in information security accountability in the cloud
10. Ali MB, Wood-Harper T, Ramlogan R (2020) A framework strategy to overcome trust issues on cloud computing adoption in higher education. In: Modern principles, practices, and algorithms for cloud security. IGI Global, pp 162–183

11. Bass C (2019) The criteria cybersecurity decision makers use to evaluate the trustworthiness of a cloud computing storage service for financial data: a qualitative study. Colorado Technical University
12. Bouzerzour NEH, Ghazouani S, Slimani Y (2020) A survey on the service interoperability in cloud computing: client-centric and provider-centric perspectives. Softw Pract Exp 50(7):1025–1060
13. Brumă LM (2020) Data security methods in cloud computing. Inf Econ 24(1)
14. Buyya R, Yeo CS, Venugopal S, Broberg J, Brandic I (2009) Cloud computing and emerging IT platforms: vision, hype, and reality for delivering computing as the 5th utility. Futur Gener Comput Syst 25(6):599–616
15. Chang H (2013) Is ISMS for financial organizations effective on their business? Math Comput Modell 58(79):79–84
16. Chang SE, Ho CB (2006) Organizational factors to the effectiveness of implementing information security management. Indus Manag Data Syst
17. Croteau A-M, Raymond L (2004) Performance outcomes of strategic and IT competencies alignment†. J Inf Technol 19(3):178–190
18. Daniel E, Vasanthi N (2019) LDAP: a lightweight deduplication and auditing protocol for secure data storage in cloud environment. Clust Comput 22(1):1247–1258
19. Dhillon G, Torkzadeh G (2006) Value-focused assessment of information system security in organizations. Inf Syst J 16(3):293–314
20. Dodge RC Jr, Carver C, Ferguson AJ (2007) Phishing for user security awareness. Comput Secur 26(1):73–80
21. Escherich M (2014) Gartner survey shows U.S. consumers have little security concern with BYOD
22. Fossi M, Egan G, Haley K, Johnson E, Mack T, Adams T, Wood P (2011) Symantec internet security threat report trends for 2010. Semant Rep 16:20
23. Gartner (2020) Gartner forecasts worldwide public cloud end-user spending to grow 18% in 2021. https://www.gartner.com/en/newsroom/press-releases/2020-11-17-gartner-forecasts-worldwide-public-cloud-end-user-spending-to-grow-18-percent-in-2021
24. Ghosh S (2020) Addressing accountability in cloud computing: a qualitative study of business cloud consumers. Wilmington University, Delaware
25. Hong KS, Chi YP, Chao LR, Tang JH (2003) An integrated system theory of information security management. Inf Manag Comput Secur
26. Ilten C, Kroener I, Neyland D, Postigo H (2012) Managing privacy through accountability. Springer
27. Ismail UM, Islam S (2020) A unified framework for cloud security transparency and audit. J Inf Secur Appl 54:102594
28. Jaatun MG, Pearson S, Gittler F, Leenes R, Niezen M (2016) Enhancing accountability in the cloud. Int J Inf Manag
29. Jaatun MG, Pearson S, Gittler F, Leenes R, Niezen M (2020) Enhancing accountability in the cloud. Int J Inf Manag 53:101498
30. Jaatun MG, Tøndel IA, Moe NB, Cruzes DS, Bernsmed K, Haugset B (2017) Accountability requirements for the cloud. In: Paper presented at the 2017 IEEE international conference on cloud computing technology and science (CloudCom)
31. Jena T, Mohanty J (2017) Cloud security and jurisdiction: need of the hour. In: Paper presented at the proceedings of the 5th international conference on frontiers in intelligent computing: theory and applications
32. Ju TL, Chen S-H, Li C-Y, Lee T-S (2005) A strategic contingency model for technology alliance. Indus Manag Data Syst 105(5):623–644
33. Kalpana P, Singaraju S (2012) Data security in cloud computing using RSA algorithm. Int J Res Comput Commun Technol IJRCCT. ISSN: 2278-5841.
34. Keeney RL, Keeney RL (2009) Value-focused thinking: a path to creative decisionmaking. Harvard University Press

35. Kelf S (2020) The security risks created by cloud migration and how to overcome them. Netw Secur 2020(4):14–16
36. Ko RK, Jagadpramana P, Mowbray M, Pearson S, Kirchberg M, Liang Q, Lee BS (2011) TrustCloud: a framework for accountability and trust in cloud computing. In: Paper presented at the 2011 IEEE world congress on services
37. Lee G, Epiphaniou G, Al-Khateeb H, Maple C (2019) Security and privacy of things: regulatory challenges and gaps for the secure integration of cyber-physical systems. In: Paper presented at the third international congress on information and communication technology, Singapore
38. Liu T, Guan X, Qu Y, Sun Y (2012) A layered classification for malicious function identification and malware detection. Concurr Comput Pract Exp 24(11):1169–1179
39. Lynn T, van der Werff L, Fox G (2020) Understanding trust and cloud computing: an integrated framework for assurance and accountability in the cloud. In: Data privacy and trust in cloud computing. Palgrave Macmillan, Cham, pp 1–20
40. Manral B, Somani G, Choo K-KR, Conti M, Gaur MS (2019) A systematic survey on cloud forensics challenges, solutions, and future directions. ACM Comput Surv (CSUR) 52(6):1–38
41. Markus ML (2004) Technochange management: using IT to drive organizational change. J Inf Technol 19(1):4–20
42. Mell P, Grance T (2011) The NIST definition of cloud computing
43. Morin J, Aubert J, Gateau B (2012) Towards cloud computing SLA risk management: issues and challenges. In: Paper presented at the system science (HICSS), 2012 45th Hawaii international conference
44. Muppala J, Shukla D, Patil S (2012) Establishing trust in public clouds'. J Inform Tech Softw Eng 2:e107
45. Mwenya JK, Brown I (2019) Cloud privacy and security issues beyond technology: championing the cause of accountability
46. Olaloye F, Adeyemo A, Edikan E, Lawal C, Ejemeyovwi J (2019) Cloud computing in education sector: an extensive review. Int J Civil Eng Technol 10:3158–3171
47. Orlikowski WJ, Gash DC (1994) Technological frames: making sense of information technology in organizations. ACM Trans Inf Syst (TOIS) 12(2):174–207
48. Pal R, Hui P (2012) Economic models for cloud service markets. In: Distributed computing and networking. Springer, pp 382–396
49. Patel P, Ranabahu AH, Sheth AP (2009) Service level agreement in cloud computing
50. Pearson S, Wainwright N (2013) An interdisciplinary approach to accountability for future internet service provision. Int J Trust Manag Comput Commun 1(1):52–72
51. Pearson S, Charlesworth A (2009) Accountability as a way forward for privacy protection in the cloud. In: IEEE international conference on cloud computing, pp 131–144
52. Pearson S (2011) Towards accountability in the cloud. In: Proceedings of the IEEE internet computing, pp 64–69
53. Pearson S (2017) Strong accountability and its contribution to trustworthy data handling in the information society. In: Paper presented at the IFIP international conference on trust management
54. Pearson S, Luna J, Reich C (2015) Improving cloud assurance and transparency through accountability mechanisms. In: Guide to security assurance for cloud computing. Springer, pp 139–169
55. Pearson S, Tountopoulos V, Catteddu D, Südholt M, Molva R, Reich C et al. (2012) Accountability for cloud and other future internet services. In: Paper presented at the 4th IEEE international conference on cloud computing technology and science proceedings
56. Potluri S, Rao KS (2020) Improved quality of service-based cloud service ranking and recommendation model. Telkomnika 18(3):1252–1258
57. Puhakainen P, Siponen M (2010) Improving employees' compliance through information systems security training: an action research study. MIS Q 34(4)
58. Purnaye P, Kulkarni V (2021) A comprehensive study of cloud forensics. Arch Comput Methods Eng 1–14.

59. Rashid ZN, Zeebaree SR, Shengul A (2019) Design and analysis of proposed remote control-ling distributed parallel computing system over the cloud. In: Paper presented at the 2019 international conference on advanced science and engineering (ICOASE)
60. Raza MR, Varol A (2020) QoS parameters for viable SLA in cloud. In: Paper presented at the 2020 8th international symposium on digital forensics and security (ISDFS); The Best Practices Act of 2010 and Other Privacy Legislation, 2010 (2010)
61. Rush B (2010) The Best Practices Act of 2010 and Other Privacy Legislation, 2010
62. Ryan P, Crane M, Brennan R (2020) Design challenges for GDPR RegTech. arXiv preprint arXiv:2005.12138
63. Saravanan N, Mahendiran A, Subramanian NV, Sairam N (2012) An implementation of RSA algorithm in google cloud using cloud SQL
64. Segev A, Porra J, Roldan M (1998) Internet security and the case of Bank of America. Commun ACM 41(10):81–87
65. Shetty J, Babu BS, Shobha G (2020) Proactive cloud service assurance framework for fault remediation in cloud environment. Int J Electr Comput Eng 10(1):2088–8708
66. Singh HP, Singh R, Singh V (2020) Cloud computing security issues, challenges and solutions (2516-2314)
67. Sreenivas V, ArunaKumari B, VenkataRao J (2012) Enhancing the security for information with virtual data centers in cloud. In: Future wireless networks and information systems. Springer, pp 277–282
68. Sreenivas V, Narasimham C, Subrahmanyam K, Yellamma P (2013) Performance evaluation of encryption techniques and uploading of encrypted data in cloud. In: Paper presented at the 2013 fourth international conference on computing, communications and networking technologies (ICCCNT).
69. Straub DW, Welke RJ (1998) Coping with systems risk: security planning models for management decision making. MIS Q 441–469
70. Subashini S, Kavitha V (2011) A survey on security issues in service delivery models of cloud computing. J Netw Comput Appl 34(1):1–11
71. Sun P (2020) Security and privacy protection in cloud computing: discussions and challenges. J Netw Comput Appl 160:102642
72. Tabrizchi H, Rafsanjani MK (2020) A survey on security challenges in cloud computing: issues, threats, and solutions. J Supercomput 76(12):9493–9532
73. Takabi H, Joshi JB, Ahn G-J (2010) Security and privacy challenges in cloud computing environments. In: IEEE security & privacy, no 6, pp 24–31
74. Tan FB, Hunter MG (2002) The repertory grid technique: a method for the study of cognition in information systems. MIS Q 26(1); Data Protection Accountability: The Essential Elements (2009a)
75. The Centre for Information Policy Leadership (2009a) Data protection accountability: the essential elements, Hunton & Williams LLP, US
76. The Centre for Information Policy Leadership (2009b) Galway project plenary session introduction, US
77. The Centre for Information Policy Leadership (2010) Demonstrating and measuring account-ability – the accountability project – Phase II Paris, France, France
78. The Centre for Information Policy Leadership (2011a) Getting accountability right with a privacy management program, Hunton & Williams LLP, Washington, DC
79. The Centre for Information Policy Leadership (2011b) Implementing accountability in the marketplace a discussion document accountability Phase III - The Madrid Project, Hunton & Williams LLP, Madrid
80. The Centre for Information Policy Leadership T. (2020) Are our privacy laws asking too much of consumers and too little of businesses? http://www.informationpolicycentre.com/2/post/2019/12/are-our-privacy-laws-asking-too-much-of-consumers-and-too-little-of-businesses.html
81. Toney SB, Kadam SU (2013) Cloud information accountability frameworks for data sharing in cloud—a review. Int J Comput Trends Technol 4(3)

82. Trompeter CM, Eloff JHP (2001) A framework for the implementation of socio-ethical controls in information security. Comput Secur 20(5):384–391
83. Vairagade RS, Vairagade NA (2012) Cloud computing data storage and security enhancement. Organization 1:2
84. Vaishnav J, Prasad N (2021) Security aspects in cloud tools and its analysis—a study. In: Inventive systems and control. Springer, pp 927–937
85. Vedder A, Naudts L (2017) Accountability for the use of algorithms in a big data environment. Int Rev Law Comput Technol 31(2):206–224
86. Vithanwattana N, Mapp G, George C (2017) Developing a comprehensive information security framework for mHealth: a detailed analysis. J Reliab Intell Environ 3(1):21–39
87. Wang C, Wang Q, Ren K, Lou W (2010) Privacy-preserving public auditing for data storage security in cloud computing. In: Paper presented at the INFOCOM, 2010 Proceedings IEEE
88. Wang Z, Yan W, Wang W (2020) Revisiting cloud migration: strategies and methods. In: Paper presented at the Journal of Physics: conference series
89. Weitzner DJ, Abelson H, Berners-Lee T, Feigenbaum J, Hendler J, Sussman GJ (2008) Information accountability. Commun ACM 51(6):82–87
90. Wong T-S, Chan G-Y, Chua F-F (2019) Adaptive preventive and remedial measures in resolving cloud quality of service violation. In: Paper presented at the 2019 international conference on information networking (ICOIN)
91. Yao J, Chen S, Wang C, Levy D, Zic J (2010) Accountability as a service for the cloud. In: Paper presented at the 2010 IEEE international conference on services computing
92. Yellamma P, Narasimham C, Sreenivas V (2013) Data security in cloud using RSA. In: Paper presented at the 2013 fourth international conference on computing, communications and networking technologies (ICCCNT)
93. YenimanYildirim E, Akalp G, Aytac S, Bayram N (2011) Factors influencing information security management in small-and medium-sized enterprises: a case study from Turkey. Int J Inf Manage 31(4):360–365
94. Zissis D, Lekkas D (2012) Addressing cloud computing security issues. Future Gener Comput Syst 28(3):583–592

Determining Vulnerabilities of Pervasive IoT Devices and Their Geographic Distribution

Segun Awoniyi and Muhammad Ali Naqi Kazmi

Abstract Projections have it that by 2023 there will be a global per capita uptick of 1.2 of networked devices which aims at hitting 29 billion mark (https://www.cisco.com/c/en/us/solutions/collateral/executive-perspe ctives/annual-internet-report/white-paper-c11-741,490.html). This is a massive expansion of the attack surface, a land flowing with milk and honey for the nefarious hackers. With review showing research focus not being favourable toward vulnerability assessment of these connected devices, especially the Internet of Things (IoT), there is a danger of pandemic of security breaches in the near future. One effective way to circumvent this looming crisis is to craft a viable framework to implement on a real-time basis the identification or detection of these devices when they join a network and measure their vulnerability with a view to either mitigate or expunge from the network altogether. Using the capabilities of Python programming, the experiment retrieves device IPs from through a network analysis tool and automatically parses into Web Application Interface (WAI) capability of Nessus vulnerability scanner to both assess and score or rank the existing vulnerabilities and severity respectively. Results showed that the device vulnerability scoring or ranking will facilitate prompt remediation decision to secure the rest of the network against any potential breach.

Keywords Network traffic analysis · Machine learning · IoT · Python programming · Nessus vulnerability scanner

1 Introduction

As the world population is currently estimated at 7.84 billion [37], Cisco Annual Internet Report (2018–2023) projects that there will be more than 29 billion networked devices by 2023 which is about 3.6 per capita of networked devices, an uptick of 1.2 from 2018 [8]. That many devices in circulation implies there will

S. Awoniyi · M. A. N. Kazmi (✉)
Northumbria University London, London, UK
e-mail: muhammad.kazmi@northumbria.ac.uk

© The Author(s), under exclusive license to Springer Nature Switzerland AG 2021
R. Montasari et al. (eds.), *Challenges in the IoT and Smart Environments*,
Advanced Sciences and Technologies for Security Applications,
https://doi.org/10.1007/978-3-030-87166-6_9

211

an astronomical increase in the data generated by the use these devices which will seamlessly expand the attack surface. The December, 2020 MacAfee report on the cost of cybercrime estimated a 50% spike since 2018, putting it at an estimate of $945 billion. The statistical graphs of cyberattacks presented by www.hackmageddon.com [20] in Figs. 1 and 2 complement this grim outlook. It can be seen that the attacks cut across all sectors of the economy and it affected individuals and enterprises alike.

Mindful of these statistics, it is appropriate to be concerned about the individual and enterprise networks that the forecasted ocean of devices will be interacting through. Therefore, formulating a framework that will identify these devices, especially the IoT, when they join any network and determining their vulnerability will facilitate the crafting of remediation strategies to protect the valuable assets from becoming victims of nefarious hackers. Otherwise, the consequence could be catastrophic. Research efforts must be re-focused on vulnerability assessment of these networked devices to pre-empt a pandemic of security violations. This is why this project will play a vital role in stemming the tide.

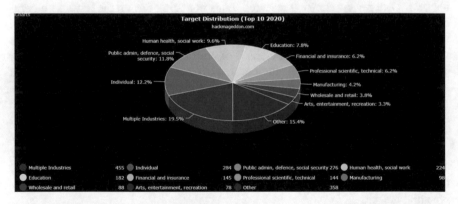

Fig. 1 Target distribution of cyberattacks [20]

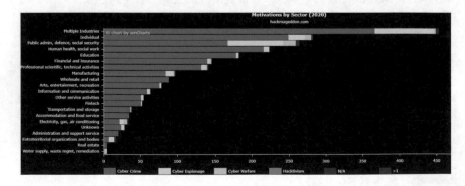

Fig. 2 Motivations of cyberattacks by sector [20]

2 Literature Review

Vulnerability is *"a flaw within a system, application or service which allows an attacker to circumvent security controls and manipulate systems in ways the developer never intended"* [29]. It is also defined as *"a weakness that allows a threat actor, such as an attacker, to perform unauthorized action (e.g., accessing the data without proper privileges"* [22, 23]. Consequently, vulnerability assessments, according to Hu et al. *"aim to test a computer system, network, or application to identify, measure, and rank vulnerabilities within the system for systematic mitigation"*. As alluded to in the Introduction chapter, it is detrimental to overlook investigation of device vulnerabilities in recognition of the fact that hackers are increasingly attacking IoT based devices. Consequently, to effectively protect the network, implementation of a security strategy must be capable of analysing and identifying critical device-specific vulnerabilities at the device level.

2.1 IoT Device Composition

As a complex interconnected system, *"the IoT ecosystem consists of a vast network of real-world objects, sensors and actuators which are connected together to form a single entity which is more intelligent and capable of interacting with the end-users, external environment and among each other"* [21].

Whether it is a large and complicated automotive manufacturing machine or a relatively smaller wearable smart bracelet, the IoT devices have in-built fixed hardware and software components such as logic chips, memory, flash storage, network module, and serial debug interface, for the hardware and boot loader and firmware for the software part [33]. At least five basic elements are required for the proper functioning of IoT devices at their deployment.

"Identification—each IoT object is identified within a network by a shareable name and a unique address. Sensing—The use of sensing devices to collect information from objects such as actuators, RFID tags, smart sensors, wearable sensing devices, etc. Communication—the ability of connected IoT devices to communicate with different devices to send and receive messages, files and other information" Burhan et al. [6].

Computation—Devices are employed to apply computation on the data collected from the objects and are used in combination with custom Operating Systems to sift unwanted information [2]

Services—Mohammed et al. and Xiaojiang et al. [30] classify IoT services into Ubiquitous services, Identity-related, Collaborative-aware, and Information Aggregation provided by a vast array of IoT applications for the development optimisation and accelerated implementation of applications.

Semantics—The incorporation of semantics has aided the creation of machine-interpretable and self-descriptive data in the IoT domain to address the challenges of interoperability between the "things" on the internet of things [5].

The aforementioned elements form the basis for varied architectural designs adopted by the manufacturers as discussed next.

2.2 IoT Architecture Considerations

As a multilayer network, the IoT interconnects with other devices to acquire, exchange, and process information. The features of each architectural layer warrant different security requirements. Therefore, the system architecture must assure operational guarantees for the IoT that links virtual worlds to the physical devices [16].

Architecture standardisation can be regarded as the foundation the IoT needs to produce an environment for companies to competitively deliver quality products. Furthermore, long-established internet architecture must be fine-tuned to address the IoT challenges. For instance, many of the underlying protocols should factor in the willingness of vast number of objects to connect to the internet [2].

The generic architectural concept has been calibrated into varying layered models by researchers over time because achieving unitary reference model has been unattainable as different organisations and users have diverse use case requirements. Nevertheless, any proposed model must be a derivative of the standardised IoT Reference Architecture composed around common vocabulary, reusable designs and industry best practices to promote interoperability between the devices and applications. Figures 3 and 4 show the generic and 3-tier IoT architectures respectively.

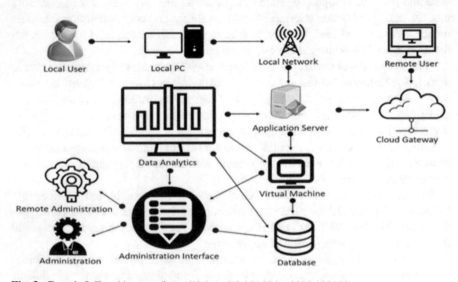

Fig. 3 Generic IoT architecture (https://doi.org/10.1016/j.iot.2020.100162)

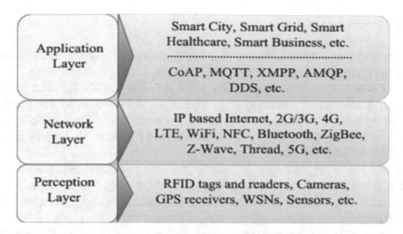

Fig. 4 3-tier IoT architecture (onlinelibrary.wiley.com/doi/full/10.1002/spy2.20)

2.2.1 Application Layer

This layer defines the applications in which the IoT can be deployed and ensures the delivery of application-specific services to the consumers [2, 6]. However, increase in the global usage of IoT applications has led to the expansion of the attack surface which has made the layer highly susceptible to hacking activities.

2.2.2 Network Layer

The network devices, servers, and smart devices are connected through the network layer. Data collected by sensors are also transmitted and processed through this layer [6]. Common to the network layer are privacy breaches, cluster security problems, compatibility issues, and traditional security problem [34].

2.2.3 Perception Layer

The sensors interact directly with the physical environment to gather information sensed about other objects in that same environment. Cvitić et al. [9] noted that some vulnerabilities are associated with the IoT architectural hierarchy and the restrictions imposed on devices at the perception layer to attain a high degree of autonomy are largely responsible for the limitations in the storage and process resources. Some of the vulnerabilities in this layer are identified as tag cloning, eavesdropping, identity theft, information leakage among others.

Researchers and manufacturers have expanded the 3-tier to 4-tier, for example, to integrate the middleware layer. Different layers are introduced to accommodate the functionalities the products are designed to perform.

2.2.4 Middleware Layer

This layer carries out processing and storage on the data obtained from the network layer and serves as the link between the system and the cloud and database.

Over time IoT and cloud computing could see advanced development that will enable immense storage and computing capabilities at this layer. In the meantime, the APIs are to meet the application layer demands. Consequently, the application layer is affected in terms of the quality of service offered by middleware layer because of the cloud and database security issues [7].

In addition to the layer-specific security concerns associated with the design of each of perception, network, and application layers, the significant non-uniformity in hardware and software makes the already difficult IoT security problems harder to tackle.

These layer-specific security requirements are prompted by layer-specific vulnerabilities [6]. The vulnerabilities attributed to different layers of the IoT architecture result in various attacks when and if exploited. Contingent upon the aims of this project which are the attempts to forestall potential DDoS and Botnet attacks, the following discussion focuses on the layer-specific vulnerabilities that result in potential DDoS and Botnet attacks. The question now is, can the presence of these devices be detected in the network? The next section attempts to address this question.

2.3   Device Detection and Identification

Although, vulnerability assessment is the main focus of this project, device identification precedes assessment. The target device to be scanned must first be identified within the network. Network Traffic Analysis (NTA) is the fundamental viable avenue of conducting the process of device detection and identification or recognition. The following presents some variants of NTA and application of some of the modern and advanced techniques employed by researchers to improve accuracy and precision.

2.3.1   Fingerprinting

Aneja et al. [3] defines device fingerprinting as *"the identification of a device without using its network or other assigned identities including IP address, Medium Access Control (MAC) address, or International Mobile Equipment Identity (IMEI) number"*. Miettinen et al. [18] presented what is termed IoT Sentinel which is designed to automate device identification by the type. The fingerprint passively captures an observed sequence of vendor-specific set-up communications triggered observes IoT devices joining a home network for the first time as well the order in which these devices interact or communicate with other objects throughout the setup process. The fingerprints are used to build classifiers for each type of device that joins the home network. A repository is created to store these fingerprints against which the fingerprint for any new device will be matched to determine the device type if there is a match. However, if the fingerprint is rejected by all the classifiers, then the device will be catalogued as a new device. In the event that an unknown device is accepted by more than one classifier, then the edit distance concept is applied to resolve the multi-class issue.

However, Shahid et al. [26] argues quite rightly that this methodology will be unhelpful if the setup process of any device has been missed for any reason and also in wanting to sniff an existing network wherein all the devices are already up and running. Furthermore, encrypted network traffic will pose another bottleneck to the effectiveness of this approach.

2.3.2 Machine Learning

The network traffic analysis (NTA) fundamentally involves analysing the network traffic generated by devices in the network for type classification, while Machine Learning is defined by [15] simply as "...*a collection of algorithms and techniques used to design systems that learn from data*". Machine Learning is a radical paradigm shift from traditional programming that processes a computer program with data to generate output to processing of data and output to generate a program or model. There are many variants to this approach.

For example, Shahid et al. [26] offers an approach based on the notion that IoT devices exhibit predictable patterns of behaviour within the network spurned by the specific tasks they perform. Although, the methodology applies a machine learning technique, the predominant activity is analysing streams of network packets the IoT device send and receive. Pre-processing of raw network data is carried out to use the inter-arrival times and size S of transmitted packets obtain the full duplex flows. The testbed activities span over seven days during which the source and destination ports and IP addresses are used to identify the full duplex flow of packets. The hypothesis factors in the sizes of the first S packets sent and received as well as the S-1 inter-arrival period of these packets. About 3300 instances were aggregated as the training set and 800 instances as test set. Python-based Machine Learning algorithms are applied to the train and test the datasets. The evaluation submission that "*most of the data points corresponding to the same IoT device lie close to each other whereas data points from different IoT devices lie far apart*" Shahid et al. [26] suggests that the work is aimed at trying to identify the type of IoT devices joining the network rather than distinguishing IoT devices from other device types.

Another variant of network traffic analysis approach is presented by Guo et al. [11] in which the methods used server names derived from DNS queries, server IP addresses, and TLS certificates bearing manufacturers information. The approach exploits the characteristic regular traffic exchange between the IoT devices and device-specific servers, aiming to identify the devices by identifying the servers they communicate with. Although, the hypothesis sounds simplistic, the methodology is multifarious. The work purports that the amalgamation of these three methods offers a more complete extensive coverage of IoT wider distribution.

2.3.3 Deep Learning

LeCun et al. [14] defines Deep Learning as allowing "...*computational models that are composed of multiple processing layers to learn representations of data with multiple levels of abstraction*". Kotak et al. [13] presented an application of deep learning to automate the identification of IoT devices connected to the network by representing the communication behaviour of these devices in small images. The work attempts to eliminate manual input that must define features from the datasets before machine learning application. The approach focuses on the TCP content, that is, the payload of the TCP packets exchanged between devices rather than measuring the packet sizes or sequence of communications during a setup process.

The pre-processing procedure indicates that representation learning is the basis of the experimental approach. Using the SplitCap tool, the singular big pcap file created from the sessions network traffic is broken into numerous pcap files with each consisting groupings of packets by IP address, protocol, and port number of same source and destination belonging to a single session. The scope of the application is restricted to TCP protocol. The resultant multiple folders contain pcap files which are regrouped by the source MAC address. The source MAC address is now used to identify the folders containing the IoT device traffic that are used in the experiments. The next stage is the stripping of packet headers in order to obtain the TCP payload content which have been converted to hexadecimal format. The binary file size is adjusted to 784 bytes and optionally converted to 28×28 pixel grey scale images. It was established from the images that there are distinctive communication patterns that distinguish one IoT device type from other IoT device types. The eventual acquired dataset was allocated to into training, validation, and test. Ultimately, the modelling enabled the prediction of IoT devices in the network.

Fundamental to all of the discussed approaches and techniques for successful detection or identification are;

- First—relatively big datasets, high number of devices and long periods of monitoring and data acquisition required for the testbeds. For instance, Miettinen et al. collected traffic measurements from 27 IoT devices for fingerprinting, setup procedure was repeated 20 times per device tested. Kotak et al. used and monitored 28 unique IoT devices for 128 days, and Guo et al. 10 used IoT devices.
- Second—pre-definition of attributes or extraction of features form the primary database against which matching process will be performed. Therefore, the foreknowledge of all the IoT devices used to build a dataset is inevitable, whether the dataset is made up of a singular or zillion devices.
- Third—a matching process that compares the features or attributes of the unknown or new device in the network to the pre-defined attributes or the extracted features.

2.4 Vulnerability Management

Vulnerability management (VM) is the continuing enterprise IT process involving vulnerability assessment, measurement, and evaluation of organisational Information System and applications with the aim of protecting critical IT infrastructure against security violation [28]. This definition supports the aims of this project. Additionally, Vulnerability Management contributes to compliance with industry relevant ISO standards.

One of the key actions to implement an effective VM involves the use of vulnerability scanner in determining existing vulnerabilities and the measuring the appropriate risk status [1]. It should be noted that VM with specific reference to the use of scanner is fraught with some limitations such as reporting of false positives in every otherwise resource-consuming vulnerability scan and inability to uncover zero-day threats and exploits as well protecting against the vulnerabilities detected. The time is right to discuss vulnerability analysis.

2.4.1 Vulnerability Analysis

Vulnerability analysis of devices is a combination of discovery and detection techniques as published by [31, 35] respectively.

Yu et al. [33] distinguishes between discovery and detection as the mining of unknown vulnerabilities and detecting the presence already existing known vulnerabilities respectively but also argue that because the dividing line between these two techniques is very blurred, researchers are increasingly concentrating their efforts on vulnerability analysis of the devices because of the considerable rise in the attack trends in these devices. In discussing the challenges to effective vulnerability analysis of IoT, Yu et al. cited complication and variety of device, constraints of device resources, and intellectual property measures as main causes.

The Nessus scanning tool in this project obviously defaults to detection analysis technique because of the general characteristics of vulnerability scanners previously highlighted.

2.4.2 Vulnerability Assessment and Techniques

Williams et al. [29] states that "*vulnerability assessments aim to test a computer system, network, or application to identify, measure, and rank vulnerabilities within the system for systematic mitigation*" and to provide organisations with knowledge of systems susceptible to cyber-attacks. Since assessment is often synonymously used with scanning, then it can be defined as "*... a process that identifies and evaluates network vulnerabilities by constantly scanning and monitoring your organization's entire attack surface for risks*" [28].

Active and passive techniques are two widely accepted techniques that could be utilised to assess and identify the vulnerabilities of IoT devices. This is a very vital aspect of vulnerability assessment. Passive vulnerability assessment is harmless to the devices because it inspects different aspects of the system installation like the Operating System, picks up the version number and checks if that version exists in the list of already known vulnerabilities hosted by the National Vulnerability Database (NVD). While passive might be considered safer, it might not reflect the true state of the system exposure because it primarily adopts historically managed database [29]. Conversely, active scanning that involves direct probing of open ports might cause some damage to the device and cause disruption in operation as a result [19].

2.4.3 Vulnerability Assessment Scanners

It is a common knowledge that vulnerability scanners generate false positives as well as their incapability to detect zero-days. Hence, the need to constantly update the plugins and vulnerability databases. It is therefore necessary to ascertain accuracy of results.

Nessus–Nessus was built as a commercial and open-source cross-platform tool with deep understanding of information security practices and is designed to simplify vulnerability assessment of both software and hardware. The pre-configured scan templates and policies reduce the stress of not knowing where or how to start assessment of asset vulnerability. The GUI provides the users a rich interactive experience and a sense of being in control of the scanning activities. In addition to its speed, it offers remediation options and links to external resources, references and databases such as Common Vulnerability Exposure (CVE) database. It employs CVSS to calculate risk analysis and scoring [28]. Nessus leads others in nearly all performance metrics.

The United States government has adopted Nessus for all vulnerability management requirement in all its foreign and domestic parastatals [36]. This appraisal of Nessus has informed its preference for this project.

There are other versatile web-based scanners such as Burp Suite, OpenVAS, Nexpose, and Metasploit.

2.5 Known IoT Vulnerabilities

Having examined some of the architecture-specific vulnerability, the following are the device-level vulnerabilities that might be exploited and the corresponding attack that might cause harm to the entire system.

2.5.1 Missing Authorisation

The term missing authorisation is a bit of a misnomer. Usurped authorisation is more direct. This is a situation that a user endeavours to perform a function outside the privileges of access associate to the defined role. This can either occur due to misconfiguration of access control or of an identity theft. And once this unauthorised user is authenticated by the device, this vulnerability could be exploited. This explains why this project considered as inappropriate to hardcode login credentials into a Python program to perform scan on valuable assets. This unauthorised or usurped privileges can be potentially damaging.

2.5.2 Account Enumeration

Account enumeration can potentially lead to Mirai Botnet attack. This might occur when a seemingly harmless system response indicating whether a user login attempts fail or succeed. The hackers are able to establish when the system finally authenticates the user with valid account identifiers after a series of failed-attempt feedback.

2.5.3 Insecure Cloud Interface

Confidentiality can be breached when data is transmitted over insecure non-SSL cloud connections. This might lead to a passive attack like eavesdropping which is one of the activities of Man-in-the-Middle attack. However, determining whether the cloud interface is secure or not is a function of many other things like no account lockout [27].

2.5.4 Vulnerability Exploitation – Mirai Example

As it was alluded to before, the pervasiveness and common-place attributes of IoT devices attract relentless assault from malicious hackers with the singular aim of causing DDoS by using botnets like Mirai.

Mirai primarily zeroes in on highly vulnerable devices especially the ones that run Busybox version of any kind. It propagates its infection by attacking them first. Afterward, it is able to extract user login credentials from a dictionary that stores username-password pairs that will provide the basis for a successful brute force attack. Other devices in the network then become easy catch [12].

This underscores the aim of this project to isolate that highly vulnerable device in order to protect the rest in the network. Figure 5 above diagrammatically portrays how the deadly botnet operates.

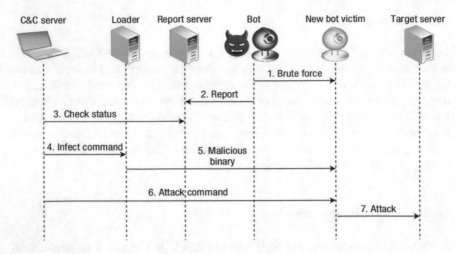

Fig. 5 Mirai botnet operation and communication (researchgate.net)

2.6 Previous Related Works

With the goal of determining and measuring the vulnerability of an IoT device in a network, previous cybersecurity academic research contributions in the academia have significantly influenced the evolution of the proposed framework.

Samtani et al. [24, 25] proposed the identification of SCADA systems and their vulnerabilities by utilising Shodan web interface to globally access and extract vulnerable devices which are processed using text mining and data mining procedure before parsing the data in a Nessus vulnerability assessment tool. This methodology enabled access to as many as 500,000 SCADA devices with many thousands classified as critically vulnerable because of default credential issues or outdated software.

Williams et al. [29] adopted the same tools as Samtani et al. to access over 150,000 consumer IoT devices, scanned them for vulnerabilities with the aims of addressing the questions of which categories of IoT device are most vulnerable and what vulner-abilities these devices are predisposed to. The research efforts were motivated by the belief that IoT vulnerabilities should be identified on a large scale by leveraging the maturity of assessment techniques.

O'Hara et al. [19] utilised combined features of Shodan and Censys for internet-wide data acquisition on pre-defined search queries on some target devices. However, the research applied an improvised tool called Scout, rather than Nessus, for vulnera-bility assessment which should conduct an active scanning without causing disruption to the target device and network.

Samtani et al. [24, 25] explored the utilisation of active and passive techniques to assess vulnerability of SCADA system. The NIST-managed the National Vulnera-bility Database (NVD) was cross-referenced for passive assessment of the devices. The proprietary Nessus scanner was used for active assessment because it directly probes the devices in search of vulnerabilities.

2.6.1 Analysis of Related Works

Resource requirements and consumption are too enormous to justify the results of these efforts. Network bandwidth will be burnt up in the process with astronomical overhead. For example, Samtani et al. [24, 25] provides the requirements as:

Master–slave architecture of 9 machines
1 master × 500 GB, 16 GB RAM, 8 CPUs
8 slaves × 80 GB SSD, 32 GB RAM, 36 CPUs.

All the cited projects provide just statistical information that can at best be useful in equipment procurement for the concerned organisation or to better understand the state of their equipment.

Real-time threats that IoT vulnerability poses to the organisational daily operations are not addressed by these research efforts. Although they all utilise the right tools, these tools ought to be linked to live operating environment and be made to handle unpredicted threats from any device in the network traffic that might not be in the statistical repository. Cumbersome data acquisition and maintenance of a repository is a very costly venture in addition to the fact that the data acquisition is historical. However, Samtani et al. [24, 25] took a more positive step to conduct active assessment by using the Nessus scanner which is intrusive in nature but provides a near perfect current state of the devices. The scanner sees the devices and the software that drives as they are. It should be highlighted that the scanner also uses the NVD in combination with other tools like CVSS to appropriately calculate the severity of the vulnerability.

2.7 The Proposed Framework

Drawing from the ideas, hypothesis and details from previous works and combining and improving on the implemented frameworks led to the construction of this proposed solution that will enhance Information System operations.

Wireshark or any other networking tool capable of capturing network packets and generating pcapng files will suffice. A Python script will extract the IP address of a joining IoT or smart device and parse it to Nessus scanner using pre-defined parameters. This will activate the Nessus scanner to assess the device for any vulnerabilities. The full view of the scan result will be presented to the user for the decision to either mitigate the threats, if any, or outrightly expunge the device from the network.

3 Practical Work

3.1 Methodology

The methodology of the practical work done will be discussed in great details in this section. The programming language, the APIs and the IDEs are discussed as well as the use of any third-party units. The requirements for the testbed, system environment, test and implementation processes are presented. The tools, techniques and selection preferences are discussed.

Some of the features of the architecture diagram of the framework in Fig. 6 were inspired by Samtani et al., O'Hare et al., and Williams et al. However, their frameworks function from outside of the system while this proposed framework functions right from within the system in which an entire Information System could potentially be exploited. Therefore, determining the vulnerability of this one device might potentially forestall a catastrophe. This framework can be implemented in personal or enterprise network to improve the confidentiality, integrity, and availability (CIA) of the network because Python is a cross-platform technology.

3.2 Scope of Implementation

Having analysed and reviewed the prior works and methodically created a framework, sufficient insight has been gained to initiate how to deviate from passive acquisition of repository of data to extracting the features of IoT the devices at the time of

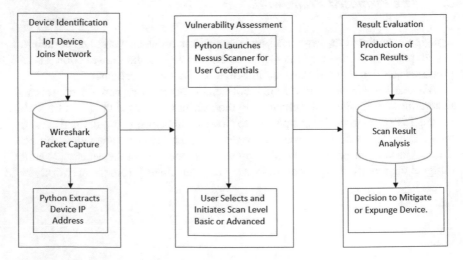

Fig. 6 High level architecture design for the proposed framework

joining a network. This is a radical departure from the method of creating a massive repository of vulnerable devices. The framework will be implemented within a home Wi-Fi network. Wireshark service will be launched and running and a smart or IoT device will be connected to the home Wi-Fi.

The sniffing tool will capture transmission packets relating to the device. A Python script that runs Pyshark module will parse live Wireshark packets to extract the relevant features of the device. The device IP address will become the target IP used by Nessus scanner to scan or assess the device for any existing vulnerability.

In the event that the administrator decides to classify a device as trusted, the device could be added to the already existing whitelist. Miettinen et al. [18] offers a three-way classification of these devices as strict, restricted and trusted based on some isolation criteria. The "Strict" denies the device internet access and confines communication to other like devices in a network zone termed untrusted overlay, "Restricted" channels the device communication to limited network destinations (for example, vendor's cloud service) including the devices in the untrust zone, and "Trusted" removes all restrictions from the devices to access the internet and other trusted devices in the network trusted zone.

3.3 Environment Requirements

The development and test equipment for the experiment testbed is extremely lite compared to previous researches discussed in the Literature Review chapter, which makes the implementation to be very less cumbersome and highly cost-effective.

Machine—ACER Nitro AN515-52 laptop Intel® Core™ i5-8300H, @ 2.30 GHz, 8.00 GB RAM, 64-bit OS, × 64-processor, 250 GB SSD and Intel P\RO/1000 MT Desktop (Bridge Adapter, Intel Wireless-AC 9560 160 MHz).

Tools—Windows 10 OS 64-bit, Wireshark 3.4.2 64-bit, Tenable Nessus 8.13.12057, Python 3.9 (Pycharm IDE),

Service—Virgin Broadband Wi-Fi.

IoT Device—TP-LINK—Kasa HS100 V2.2 Smart Plug, Kasa app, compatible with iOS (8 or later) and Android (Jelly Bean 4.1 or later), Wi-fi-enabled, Control lights and appliances remotely using your smartphone or tablet.

3.4 Implementation of Technology

The creation of an effective framework to assess the vulnerability of IoT device within a network involves many components. Convenience, scalability, user-friendliness, storage economy as well as speed form the basis for the technology used to construct the framework. The first among the main stages for the implementation of this framework is device detection and identification. Although, the central focus of this project is to assess and determine device vulnerability, devices must first be identified. More than a few researches in recent times have dwelled solely on device detection and identification. Many advanced methodologies and state-of-the-art techniques have been proposed and applied, including Machine Learning detection technique [17] (Bohadana et al. 2017), Identification of IoT devices using deep learning [13], Fingerprinting and network analysis techniques of device detection [26, 32] and many others. Machine Learning adopts the classified and unclassified approaches in its implementation. Features of devices will be defined and applied to data sets before their pre-processing. The Machine Learning appropriation culminates in application of performance metrics to iteratively measure the success rate in the modelling of raw data and mapping to the vector space and application of Machine Learning algorithms. At the heart of Machine Learning is big data and even bigger in the implementation of Deep Learning. Deep Learning offers a more accurate and predictive result. At the heart of the amazing technologies is massive acquisition and storage of object features over time to which a target object located in the groupings. This project has no luxury of big data at its disposal.

The two methods for scanning a device in the Tenable.io Nessus are through the Web Application Interface that runs on localhost port 8834 and the API that runs in a Python IDE. The former requires the download of a Nessus installer to set up the scanner while the latter requires generating an API keys. However, both methods require an account setup with Tenable because Nessus is a licenced product. Nessus Essentials which is the free version that permits scanning of a limited number of target devices was installed for this project. An input screen is designed to accept user inputs to Detect/Identify Device, Device Vulnerability Scan, or Exit the Process. Each accepted user unput will be linked to the corresponding variable within Python and trigger the right process within the Python script.

A TP-LINK-Kasa HS100 Smart Plug is switched on to connect to Wi-fi like other devices and Wireshark will be launched to capture packets of data transmitted in the network traffic. The system will accept the following required user input from the menu.

If 'Detect/Identify IoT Device?' option is 'Y' from the menu, the Python wrapper, Pyshark, will execute pyshark.LiveCapture() method to receive the live packets capture and process to identify and extract the appropriate device IP address. The process terminates if input is 'N'. "Proceed to Vulnerability Scan" option will automatically launch the Nessus WAI scanner that requires administration credentials to authenticate the user.

If input is 'Y' to option 2, the process continues. At this stage Python will invoke the Selenium web-based automation tool which can be installed in Windows command line with a pip command if not already installed.

C:\...... > pip install selenium

Selenium works with the module webdriver-manager which if not already installed can also be installed as:

C:\... > pip install webdriver-manager

There may be the need to run 'pip list' command to see the list of packages already installed with Python on the system as all relevant packages must be imported into Python to be error-free. Successful user authentication will launch the webpage to specify parameters to create a new scan. The scan parameterisation allows user to customise scan templates and set plugin rules that define the actions to be carried out during scanning. The scanner page offers three major categories of policy templates. Discovery, wherein a simple scan is performed to discover system-wide live hosts and open ports. Vulnerabilities category allows a Basic Network Scan and many other scans of preference. This project will conduct Both Basic and Advanced scans. Compliance is the final category that offers premium compliance services such as Cloud infrastructure auditing that assures the configuration compliance of the third-party cloud services. A typical vulnerability assessment result displays the number of vulnerabilities found as well as their severity rankings that range from informational to critical. Selection of the individual entries will drill down into the details of the details of the results. For example, drilling on the entry provides a full description of the vulnerability found, risk information—Risk factor, CVSS Vector, and CVSS Base Score, as well as suggested mitigation. Weblinks of more useful resources are also on offer. The user will directly and interactively perform the actual scanning of the target device and also benefits from an aesthetic and more informative GUI-based scan feedback. All of these activities discussed will be comprehensively demonstrated in the Use Case Test Procedures section.

3.5   Use Case Test Procedures

The testing methodology adopted is an on-going iterative process, particularly during the coding phase. The whole development is divided into tasks which is reflected in the way scripting has been compartmentalised. The testing process in divided into a list of 5 use case test categories. Certain security precaution must be exercised before commencing the process such as disabling remote access to the router or Access Point to prevent access to especially a home network from a non-domiciled device.

3.5.1   Connect Target Device to Wi-Fi

The pre-installed and configured IoT target device, the HS100 Kasa smart plug should be connected to Wi-Fi to get allocated IP address by Dynamic Host Control Protocol (DHCP). It is allocated 192.168.0.31 as shown in Fig. 15.

The IP address will be used to identify the device in the network (Fig. 7).

Device name	MAC address	IP address	Speed (Mbps)	Connected to
HS100	B0:95:75:48:91:DB	192.168.0.31/24	28.90	WI-FI 2.4G VM3486747

Fig. 7 Allocation of IP address to target device by router

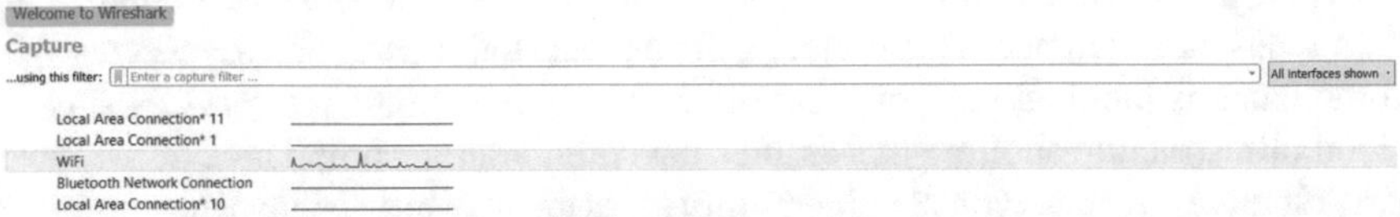

Fig. 8 Selection of Wi-Fi interface in Wireshark for packet transmission

3.5.2 Initiate Monitoring of Network Traffics

Wireshark traffic monitoring tool is launched and the Wi-Fi interface is selected (Fig. 8).

Now Wireshark is up and running as shown in Fig. 16 to capture network packets.

3.5.3 Initiate Packets Transmission by Target Device

Although, IoT devices are known to communicate with their remote servers intermittently, the transmission time cannot not be determined (Fig. 9). Therefore, transmission of packets is forced by issuing 'ping' command from the host machine initiate

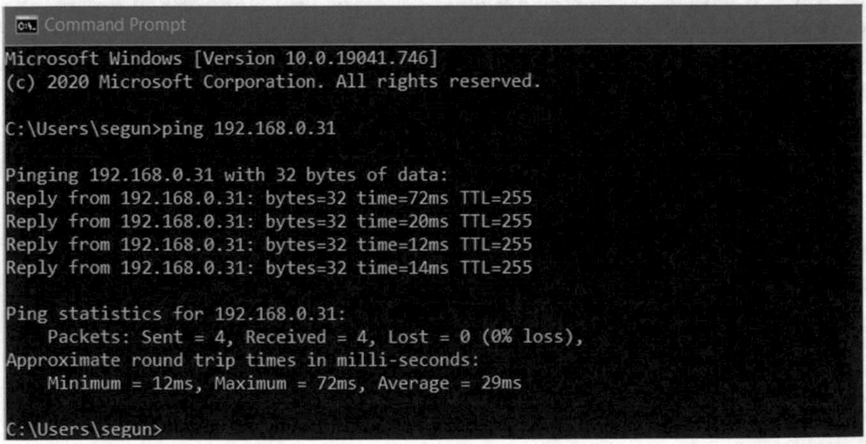

Fig. 9 Communication and packet transmission from 192.168.0.31

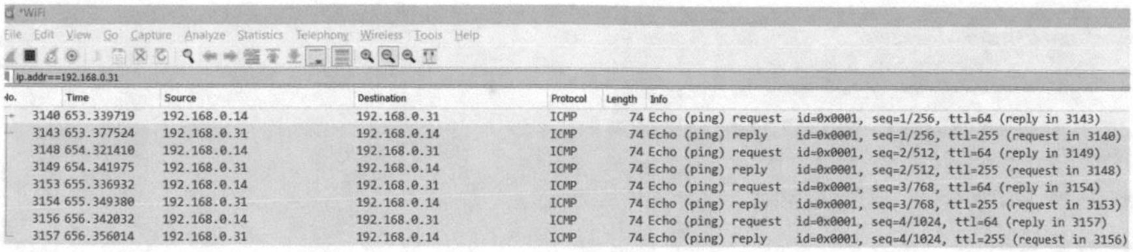

No.	Time	Source	Destination	Protocol	Length	Info
3140	653.339719	192.168.0.14	192.168.0.31	ICMP	74	Echo (ping) request id=0x0001, seq=1/256, ttl=64 (reply in 3143)
3143	653.377524	192.168.0.31	192.168.0.14	ICMP	74	Echo (ping) reply id=0x0001, seq=1/256, ttl=255 (request in 3140)
3148	654.321410	192.168.0.14	192.168.0.31	ICMP	74	Echo (ping) request id=0x0001, seq=2/512, ttl=64 (reply in 3149)
3149	654.341975	192.168.0.31	192.168.0.14	ICMP	74	Echo (ping) reply id=0x0001, seq=2/512, ttl=255 (request in 3148)
3153	655.336932	192.168.0.14	192.168.0.31	ICMP	74	Echo (ping) request id=0x0001, seq=3/768, ttl=64 (reply in 3154)
3154	655.349380	192.168.0.31	192.168.0.14	ICMP	74	Echo (ping) reply id=0x0001, seq=3/768, ttl=255 (request in 3153)
3156	656.342032	192.168.0.14	192.168.0.31	ICMP	74	Echo (ping) request id=0x0001, seq=4/1024, ttl=64 (reply in 3157)
3157	656.356014	192.168.0.31	192.168.0.14	ICMP	74	Echo (ping) reply id=0x0001, seq=4/1024, ttl=255 (request in 3156)

Fig. 10 Packet transmission from 192.168.0.31 on Wireshark

communication with the smart plug. Figure 17 confirms communication between the two devices.

Wireshark is checked in Fig. 18 to confirm packets transmission from 192.168.0.31 before launching Python program (Fig. 10).

3.5.4 Launch Python Program

Launch the Python application as 'nessusProgram.py' in any Operating System platform. This brings up the menu designed for the scanning session (Figs. 11 and 12). A 'y' or 'Y' input response to option '1' initiate a live packet capture. This has been set to maximum of 200 packets setting the interface type to correspond to Wireshark setting in Fig. 16.

Machine Learning approach is building a dataset within Python that consists of a list of extracted certain features attributable to some IoT device types that are already known to the network, against which the features of any unknown device

```
detect_iot = input('1. Detect/Identify IoT Device? (Y/N): ').upper()

if detect_iot != 'Y':
    exit()
else:
    capture = pyshark.LiveCapture(interface='WiFi')
    capture.sniff(packet_count=200)
```

Fig. 11 Python code segment for live packet capture

```
C:\Users\segun\AppData\Local\Programs\Python\Python39\python.exe C:/Users/segun/PycharmProjects/mastersProject/nessusProgram.py
VULNERABILITY SCANNING SESSION
------------------------------
1. Detect/Identify IoT Device? (Y/N): y

Device found with IP Address:  192.168.0.31
This is the Target IP for the NESSUS Scanner

2. Proceed to Vulnerability Scan? (Y/N):
```

Fig. 12 Matching of unknown devices in network against known features

```
C:\Users\segun\AppData\Local\Programs\Python\Python39\python.exe C:/Users/segun/PycharmProjects/mastersProject/nessusProgram.py
VULNERABILITY SCANNING SESSION
-------------------------------
1. Detect/Identify IoT Device? (Y/N): y

Device found with IP Address:  192.168.0.31
This is the Target IP for the NESSUS Scanner

2. Proceed to Vulnerability Scan? (Y/N): y
```

Fig. 13 Identified IoT device in the network

```
driver: WebDriver = webdriver.Chrome(executable_path="C:/Users/segun/website_login/chromedriver.exe")
driver.maximize_window()

url = 'https://localhost:8834/#/'
driver.get(url)
driver.find_element_by_id('details-button').click()
driver.find_element_by_id('proceed-link').click()
```

Fig. 14 Selenium methods to launch Nessus authentication

will be matched to determine the type. This is in line with the ML philosophy of processing an input data with output dataset to generate a model or a program. The principal features are the dynamically allocated IPs allocated to devices within a LAN or WAN which usually remains constant in the DHCP table. As has been previously explained, this project cannot afford the cost of collection of so many IoT devices as the organisationally or institutionally sponsored researches. However, the same philosophy has been adopted at the micro level. The most important takeaway is the scalability of the approach adopted by this project. The program terminates if there is no match (Fig. 13).

Figure 20 reflects the decision to connect to Nessus scanner Web Application Interface (WAI) for vulnerability assessment (Fig. 14). At this point the pre-configured Python library, Selenium, is triggered to automatically launch the scanner outside of Python which presents the user authentication session as seen in Figs. 21 and 22, haven responded successfully to security exceptions.

3.5.5 Nessus Scanner User Authentication

The user authentication is required to access the scanner functionalities. Access to the scanner is contingent upon prior subscriptions that prescribe the levels of access (Fig. 15).

The 'Not secure' message displayed on the Chrome webpage is normal. It is because the scanner runs on localhost on port 8834 and tries to validate the SSL certificate as a remote client (Fig. 16). The security exceptions flagged have been handled automatically within Python code by defaulting values to the appropriate web elements without user involvement. The SSL certificate is confirmed as valid by Fig. 23.

Fig. 15 Nessus scanner user authentication session

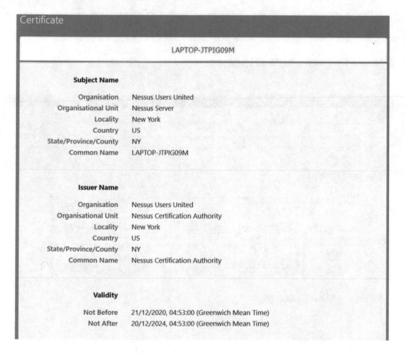

Fig. 16 Nessus SSL certificate validation

3.5.6 Launch Nessus Scanning Session

The authentication is successful and the scanning session is initiated wherein basic scan parameters are defined for an Advanced scan on the target device using the device IP previously identified in Fig. 20. A few steps are taken to prepare the session for scanning (Figs. 17, 18 and 19).

Define scan parameters.

The 'New Scan' was pressed and the 'Scan Template' page is loaded where 'Advanced Scan' was selected as seen in Figs. 24 and 25.

The 'General' option is selected and arbitrary entries are made into 'Name' and 'Description' fields and a valid target IP (192.168.0.31) is entered into the 'Targets' field. The 'Save' button is pressed to save the scan parameters as depicted in Fig. 20. The system presents the page in Fig. 20 for launching the scanning.

The small play button is pressed to launch scanning as shown in Fig. 20. This completes the use case test procedure. The obtained are obtained for analysis and discussion in the Results, Analysis and Evaluation chapter.

Fig. 17 Nessus New scan selection page

Fig. 18 Scan template selection page

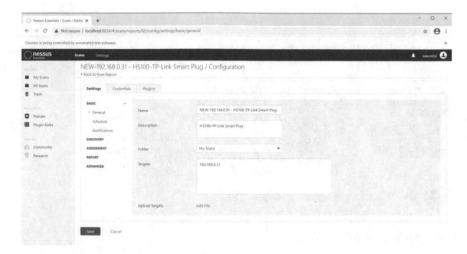

Fig. 19 Definition of scan parameters for target device

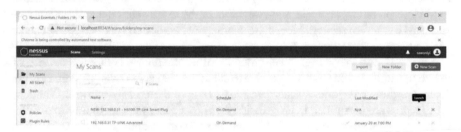

Fig. 20 Launching of device scanning

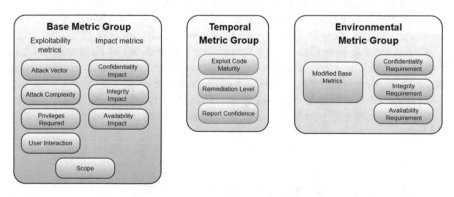

Fig. 21 Composition of vulnerability metrics (Babix.com, 2021)

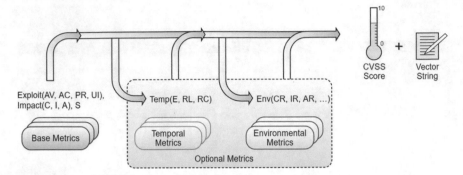

Fig. 22 CVSS metrics and equations template [10]

3.6 Vulnerability Metrics

A good understanding of vulnerability metrics will serve a useful precursor to meaningful interpretation, analysis, and evaluation of the results later in the Results Analysis and Discussion section. Metric and scoring are two of the prominent components any vulnerability assessment tool.

Figure 21 provides a high-level grouping of all the metric components. Each of the groups operates on a set of formulas to calculate the scores relating to that group. There are three categories of metrics; Base, Temporal, and Environment. Vulnerability scores are performed in progression from Basic to Temporal to Environmental in an inter-dependent format. That is, the Base Score is applied in the calculation of the Temporal Score while the Temporal Score is applied in the calculating the Environmental Score (CVSS Base Score Explained [4].

This is the underlying principle adopted by Nessus assessment tool for calculating the scores that place any target device into a severity category (Fig. 22).

All the scoring stems from Base Score which is made up of Exploitability and Impact metrics. These two metrics are very important in the entire calculation process. However, particular attention should be paid to the Impact metric as the measurement of the consequence of a successful exploit. This directly affects the very reference model for any security considerations, the confidentiality, integrity, and availability (CIA).

4 Results, Analysis and Evaluation

4.1 Results Analysis and Discussion

Although, the results of the use case testing mostly produced the expected outcome at each stage, the challenge remains that no tool or any of the frameworks examined offers any solution to the real-time network traffic flow threats. The fact that the central theme of this project is vulnerability determination and measurement does not preclude device identification in a real-time approach. There must be a uniqueness to defining the features any joining device for accurate identification. The closest option is the device Media Access Control (MAC) address that only be used to derive the disclosure of the vendor but nothing more as shown in Fig. 23.

Another challenge is the indeterminate timing of communication between IoT devices and their remote servers. Because there can be no packet transmission by these devices without communication (Fig. 24). Packet transmission is the only way these devices can be visible to Wireshark used as the monitoring tool in the project. Hence, the induced generation of traffic from the target device. Although, a scanning tool like Nmap will not necessarily wait for device-server communication from the target device, Fig. 23 confirms that it also uses 'pinging' in probing of different aspect of the device during Address Resolution Protocol (ARP) phase of the Nmap intensive scan performed on the target device.

Figure 25 is the page displaying the summary scan result at completion. Incidentally the target device is clean, although, a highly vulnerable outcome would have been preferred that will present the opportunity to see all the measurement metrics on display. However, it must be pointed out that this 'clean' result is the outcome of all the metrics and risk analysis. In general, the quick interpretation of this result is that the device has a clean bill of health and should interact with other secure devices in the network. It can also be added to the 'Trusted' category as previously discussed in the Scope of Implementation section. Drilling down on this summary reveals more details about the device as seen in Figs. 26, 27, and 28.

```
PORT     STATE SERVICE VERSION
9999/tcp open  abyss?
MAC Address: B0:95:75:48:91:DB (Tp-link Technologies)
No exact OS matches for host (If you know what OS is running on it, see https://nmap.org/submit/ ).
```

Fig. 23 Nmap device vendor identification

```
C:\Users\segun>nmap -T4 -A -v -Pn 192.168.0.31
Host discovery disabled (-Pn). All addresses will be marked 'up' and scan times will be slower.
Starting Nmap 7.91 ( https://nmap.org ) at 2021-02-04 19:09 GMT Standard Time
NSE: Loaded 153 scripts for scanning.
NSE: Script Pre-scanning.
Initiating NSE at 19:09
Completed NSE at 19:09, 0.00s elapsed
Initiating NSE at 19:09
Completed NSE at 19:09, 0.00s elapsed
Initiating NSE at 19:09
Completed NSE at 19:09, 0.00s elapsed
Initiating ARP Ping Scan at 19:09
Scanning 192.168.0.31 [1 port]
Completed ARP Ping Scan at 19:09, 0.31s elapsed (1 total hosts)
Initiating Parallel DNS resolution of 1 host. at 19:09
```

Fig. 24 Nmap ARP ping scan

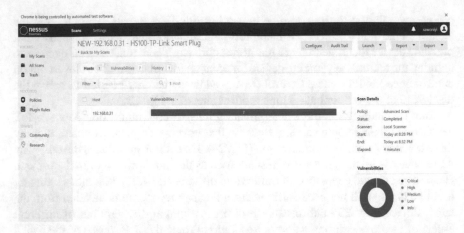

Fig. 25 Result of the advanced scan of 192.168.0.31

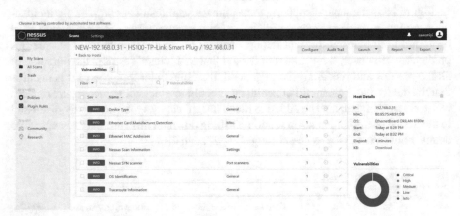

Fig. 26 First level drilldown of scan vulnerabilities

Fig. 27 Drilldown details about device type

Fig. 28 Drilldown details about device manufacturer

The first place to inspect is the right-hand side where there is a colour-coded pie chart that categorises the vulnerability of the device. In this case, it is all blue, which implies that it has only informational feedback for the just completed assessment.

Secondly, clicking on the vulnerability button drills down to see more details where more sensitive detail information about the device like the MAC address, is displayed. The type of scan and the lapsed time are given as well.

The pie chart is the first alert centre for any critical feedback from the scan. The risk level here is absolutely zero.

Two other household items were scanned to afford the chance of displaying the metric, although they are not classified as IoT. Anyway, a device is a device it is the embedded technology that changes their behaviour and performance (Fig. 29).

One of the target IP's is a smart phone and the other a gateway. At a glance, the colour-coded chart provides mixed alerts of medium vulnerability on both devices which if drilled down will provide more details and the scoring that was discussed under Vulnerability Metrics. 18 Medium vulnerabilities on 192.168.0.16 and 51 informational. The granular report for 192.168.0.16 on SSL certificate vulnerability

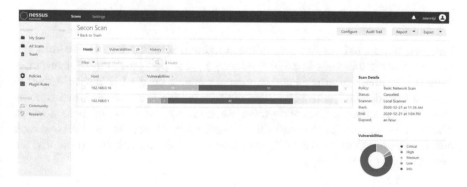

Fig. 29 Result of basic network scan on two hosts

Fig. 30 Granular scan result for 192.168.0.16 SSL certificate

is shown in Fig. 30. The Severity and Risk Information as well as Scoring are on full display. The CVSS Base score has been calculated twice using different Vectors that produced Base Scores of 6.5 and 6.4 which are virtually the same. Full description of vulnerability, suggested remediation and links to external references.

4.2   Project Evaluation

This chapter holistically evaluates the standard of implementation of the project as it is measured against the aims and objectives set out at the beginning. This discussion encapsulates the essential features of the dissertation such as the literature review, framework design and implementation as well the effectiveness of the techniques and the testing process adopted that produced the achieved results.

4.2.1   Literature Review Phase

This was a very tedious exercise because the underpinning idea is two-fold and all the available materials assembled addressed each part separately. For instance, there is a plethora of articles and academic projects on the IoT device detection and identification and quite a sizeable number of research efforts on addressing the characteristic vulnerabilities and proneness to security violation of these devices. Therefore, searching for the works that combined these tow aspects into one took almost forever and to no to avail. Too many of the available materials presented the same thing in different ways.

Although, the related works reviewed offered so little about the programmatical approach to the efforts, concentrated study of the implemented frameworks eventually yielded an innovative and novel approach adopted in this project. Also, the most prominent and useful of the researches had institutional or state sponsorship which enabled deployment of vast resources to achievement their objectives. Therefore, it became even harder to construct a much less prodigal and effective framework. The bottlenecks at this stage led to determination and motivation to forge a way rather than demoralisation. The literature review can be deemed rewarding having yielded a workable framework and expansion of knowledgebase.

4.2.2 Design Phase

The design phase became easier because of the prolonged brainstorming in the literature review phase coupled with the fact that there was very clear thinking of what to propose which was merging two propositions into one and just finding a way of implementing the new design. The design was also driven by the requirement of knowledge acquisition, timeline, and availability of resources. In the end, the design only led to modification of the methodology but the overall aims and objectives of the project remained constant. The cosmetic had to make way for the essential. The initial concept to provide geolocation heatmap of vulnerable devices was considered to be cosmetic and non-value-added to the main aim of the project. On the economy of scale, the design can be considered to be effective as well as scalable.

4.2.3 Implementation Phase

The implementation of the framework was reconsidered for some cogent reasons. Firstly, performing internal scanning as prescribed by Tenable was found to be an added vulnerability. Using that approach, a Python API offers a programmatic configuration and parameterisation of the scanner. However, login credentials (username and password) must be hardcoded within the Python script. Therefore, this project implements compulsorily incorporated user authentication. Secondly, many of the API parameters are ambiguously defined such that repetitive changes only led to crashes and timewasting. Thirdly, the scan results if successful can only be imported in formats (csv, pdf, etc.) that hampers an interactive analysis such as the ability to drilldown some of the components to better understand the severity of vulnerability of the target devices.

4.2.4 Testing Phase

The testing approach was both manual and iterative. Every line of code was tested to see the impact. Tests were carried out in two main areas. Firstly, performing a live network packets capture using Python and parsing the captured packets to extract the appropriate device IP was not as straightforward as initially assumed. Having to gain the requisite Python programming knowledge to achieve this was very time consuming which created a real challenge in the literature part of the project. However, the use case was quite simplistic because it is menu-driven.

The second stage was the decision to continue to scanning wherein the user proceed to scanner with just a button click. The user is not required to configure or define scan parameters each time a scan session is initiated. The results can clearly be regarded as a crystallisation of the main aim of the project which is to determine the vulnerability of an IoT device found in a network traffic. The assessment results will facilitate the isolation of the device for remediation or outright expungement from the network before they can be exploited to induce DDoS of the entire system.

4.2.5 Personal Benefits

The painstaking efforts at the early stages of the project have now translated to intellectual curiosity and a revived hunger to expand the knowledgebase, especially with the discovery of highly potent Python libraries. Selenium is a very powerful Python wrapper designed to automated web login and interaction with the web elements. Hours expended in consuming tutorial YouTube videos. There is definitely the drive to resume intellectual exploration after the project.

5 Conclusion and Recommendation

The pervasiveness and ubiquity of embedded wireless sensors and actuators collectively tagged IoT devices coupled with the inherent vulnerabilities in the webspace have made it absolutely critical to detect and identify the presence of these devices and critically gauge the vulnerability. However, the existing modelling has concentrated mainly on creating repositories of vulnerable devices which must be maintained and updated periodically with new vulnerable devices. This thesis presents a practicable framework constructed and implemented to forestall the potential DDoS devastation. The approach extracted some features of a joining IoT device from Wireshark packet capture using Python. The purpose was to identify these devices when as soon as they join the network. However, the process of identifying device type in the network could not be automated due to the reason stated in the Results, Analysis and Evaluation chapter.

It can be deduced and concluded from the whole exercise that organisations must commit to high premium investment strategy in implementing pre-emptive and proactive information systems security strategy subject to constant review in order to guarantee high reputation and profitability in the highly competitive and increasingly hostile business domain.

5.1 Recommendation and Future Direction

One of the areas that should be expanded upon based on the implemented framework is the automated identification of device type in a network as there is presently no known tool that can achieve that on a real-time basis. Although, some of the related works investigated tried to improvise, they were predicated on historical and passive analysis. The partial automation of web login to Nessus scanner that this project implemented could also be further researched so that after the mandatory user authentication session, Python or any other tool can pick up the processes to atomate the creation of a new scan session based on user decision incorporated into the Scan Menu. Selenium, one of the Python open-source wrappers used in the program is one of the very useful tools for web-based automation. It was used in this project to automatically launch the Nessus scanner and to automatically populate values for the security prompts preceding the user authentication. At the moment, it provides automation for the vulnerability scanning process that does not extend beyond the user authentication stage. This provides an exciting future prospect of full automation in which case user authentication leads to automatic creation of a new scan session within the Nessus scanner, selection of the appropriate scan template, defaulting some parameters as depicted in Table 1, and finally launching the scanner. This

Table 1 Nessus scaning parameters.

Operation type	System activity description	Expected user input	Data type
Obtain target device IP	Detect/identify device?	"Y" or "N"	String
	Proceed to vulnerability scan?	"Y" or "N"	String
Device scanner user authentication	Login name	Username	String
	Password	Password	String
Nessus scanner parameters	Create new scan	Click button	Not applicable
	Select scan template	Click 'Basic Network Scan' or 'Advanced Network Scan'	Not applicable
	Settings	Click 'General'	Not applicable
	Name (scan name)	Enter any name for scan	Not applicable
	Description	Enter any description for scan	Not applicable
	Target (IP of the device to scan)	Enter 'IP' address target device	
	Save parameters	Click 'Save' button	
	Run scan	Click 'Run' button	

will undoubtedly enrich the user experience, eliminate human errors, and enhance productivity that will directly or indirectly increase organisation profitability.

References

1. https://www.akamai.com/uk/en/resources/vulnerability-management.jsp 2021
2. Al-Fuqaha A, Guizani M, Mohammadi M (2015) Internet of things: a survey on enabling technologies, protocols, and applications—IEEE Journals & Magazine. [online] Ieeexplore.ieee.org. https://ieeexplore.ieee.org/document/7123563/
3. Aneja S, Aneja N, Islam S (2018) IoT device fingerprint using deep learning. [online] Arxiv.org. https://arxiv.org/pdf/1902.01926
4. Balbix (2021) CVSS base score explained. Balbix. [online] https://www.balbix.com/insights/base-cvss-scores/
5. Barnaghi P, Wang W, Henson C, Taylor K (2012) Semantics for the internet of things. Int J Semant Web Inf Syst 8(1):1–21
6. Burhan M, Rehman, Khan B (2021) https://www.researchgate.net/publication/327272757_IoT_Elements_Layered_Architectures_and_Security_Issues_A_Comprehensive_Survey
7. Chen K, Zhang S, Li Z, Zhang Y, Deng Q, Ray S, Jin Y (2018) Internet-of-things security and vulnerabilities: taxonomy, challenges, and practice. [online] Semanticscholar.org. https://www.semanticscholar.org/paper/Internet-of-Things-Security-and-Vulnerabilities%3A-Chen-Zhang/da7f58eef3aeb6283aea13e7c18175156669454f
8. Cisco (2021) Cisco annual internet report—Cisco annual internet report (2018–2023) White Paper. https://www.cisco.com/c/en/us/solutions/collateral/executive-perspectives/annual-internet-report/white-paper-c11-741490.html
9. Cvitić I, Vujić M, Husnjak S (2015) [online] Daaam.info. https://daaam.info/Downloads/Pdfs/proceedings/proceedings_2015/102.pdf
10. FIRST—Forum of Incident Response and Security Teams (2021) CVSS V3.0 specification document. https://www.first.org/cvss/v3.0/specification-document
11. Guo H, Heidemann J (2019) [online] Isi.edu. https://www.isi.edu/publications/trpublic/pdfs/isi-tr-726b.pdf
12. Kolia C, Kambourakis G, Stavrou A, Voas J (2017) DDoS in the IoT: Mirai and other botnets. [online] Ieeexplore.ieee.org. https://ieeexplore.ieee.org/document/7971869
13. Kotak J, Elovici Y (2021) Iot device identification using deep learning
14. LeCun Y, Bengio Y, Hinton G (2015) Deep learning. https://s3.us-east-2.amazonaws.com/hkg-website-assets/static/pages/files/DeepLearning.pdf
15. Lee W (2019) Python® machine learning. https://learning.oreilly.com/library/view/python-machine-learning/9781119545637/
16. Li S, Tryfonas T, Li H (2016) The internet of things: a security point of view. [online] Semanticscholar.org. https://www.semanticscholar.org/paper/The-Internet-of-Things%3A-a-security-point-of-view-Li-Tryfonas/54497abbdd4b5f18541a90fc50d215ad6d14db65
17. Meidan Y, Bohadana M, Shabta A (2017) https://www.researchgate.net/publication/319736005_Detection_of_Unauthorized_IoT_Devices_Using_Machine_Learning_Techniques
18. Miettinen M, Marchal S, Hafeez I (2016) Iot SENTINEL: automated device-type identification for security enforcement in Iot—IEEE Conference Publication. [online] Ieeexplore.ieee.org. https://ieeexplore.ieee.org/abstract/document/7980167/
19. O'Hara J, Macfarlane R, Lo O (2019) Identifying vulnerabilities using internet-wide scanning data. [online] Ieeexplore.ieee.org. https://ieeexplore.ieee.org/document/8688018
20. Passeri P (2021) 2020 Cyber attacks statistics. [online] HACKMAGEDDON. https://www.hackmageddon.com/2021/01/13/2020-cyber-attacks-statistics/
21. Peng S, Pal S, Huang L (2020) Principles of internet of things Iot ecosystem. [S.L.]: Springer

22. Rizvi S, Orr R, Cox A, Ashokkumar P, Rizvi M (2021) Identifying the attack surface for Iot network
23. Rizvi S, Pipettiz R, Mcintyres N, Todd J (2021) (PDF) Threat model for securing internet of things (Iot) network at device-level. [online] ResearchGate. https://www.researchgate.net/pub lication/341884275_Threat_Model_for_Securing_Internet_of_Things_IoT_Network_at_D evice-Level
24. Samtani S (2018) CSDL|IEEE computer society. [online] Computer.org. https://www.com puter.org/csdl/magazine/ex/2018/02/mex2018020063/13rRUyft7zb
25. Samtani S, Yu S, Zhu H, Patton M, Matherly J, Chen H (2018) Identifying SCADA systems and their vulnerabilities on the internet of things: a text-mining approach. IEEE Intell Syst 33(2):63–73 (SANS Institute: Reading Room—Threats/Vulnerabilities, 2021)https://doi.org/10.1109/MIS.2018.111145022
26. Shahid M, Blanc G, Zhang Z, Debar H (2018) IoT devices recognition through network traffic analysis. [online] Academia.edu. https://www.academia.edu/38420293/IoT_Devices_Recogni tion_Through_Network_Traffic_Analysis
27. Suo H, Wan J, Zou C, Liu J (2012) Security in the internet of things: a review—IEEE Conference Publication. [online] Ieeexplore.ieee.org. https://ieeexplore.ieee.org/document/6188257
28. Tenable® (2020) Nessus professional. https://www.tenable.com/products/nessus/nessus-profes sional
29. Williams R, McMahon E, Samtani S (2017) Identifying vulnerabilities of consumer internet of things (Iot) devices: a scalable approach—IEEE Conference Publication. [online] Ieeex plore.ieee.org. https://ieeexplore.ieee.org/document/8004904.
30. Xiaojiang X, Jian-li W, Ming-dong L (2021) Services and key technologies of the internet of things. [online] Semanticscholar.org. https://www.semanticscholar.org/paper/Services-and-Key-Technologies-of-the-Internet-of-Xiaojiang-Jian-li/3233bc706e9c0a1757de889ee5dfc1 9b0b358b32
31. Xie W, Jiang Y, Ding N (2017) CSDL|IEEE computer society. [online] Computer.org. https://www.computer.org/csdl/proceedings-article/icpads/2017/212901a769/12OmNzZWbxb
32. Yang K, Qiang L, Limin S (2019) http://static.tongtianta.site/paper_pdf/35365316-adba-11e9-b197-00163e08bb86.pdf
33. Yu M, Zhuge J, Cao M, Jian L (2020) [online] Mdpi.com. https://www.mdpi.com/1999-5903/12/2/27/pdf
34. Zhao K, Ge L (2013) A survey on the internet of things security. In: Proceedings of the 2013 ninth international conference on computational intelligence and security. [online] Dl.acm.org. https://dl.acm.org/doi/10.1109/CIS.2013.145
35. Zheng Y, Wen H, Cheng K, Song ZW, Zhu HS, Sun LM (2019) A survey of IoT device vulnerability mining techniques
36. WonderHowTo (2016) Hack Like a pro: how to scan for vulnerabilities with Nessus. https://null-byte.wonderhowto.com/how-to/hack-like-pro-scan-for-vulnerabilities-with-nessus-0169971/
37. Worldometers.info (2021) World population clock: 7.8 billion people (2021)—Worldometer. https://www.worldometers.info/world-population/

Holistic Authentication Framework for Virtual Agents; UK Banking Industry

Hasitha Hettiarachchi Hettiarachchige and Hamid Jahankhani

Abstract The exponential growth and advancements of information technology and artificial intelligence significantly enhance the capabilities of virtual agents. This creates opportunities for businesses such as banks to adapt virtual agents for banking operations. Security and privacy are the ultimate concern for any e-banking system or application. Therefore, virtual agents inherit this concern. The previous literature lacks the knowledge of a suitable authentication framework for the virtual agents use in the UK banking industry. This forms a vulnerability for the banks while providing an opportunity for cyber attackers to attack the endpoints of the virtual agents. Moreover, it produces the risk of exposing sensitive customer data. Therefore, this research was conducted to fill the knowledge gap by proposing an authentication framework for the banking virtual agents. The proposed framework considers several security standards and controls to ensure that the proposed framework fits for the purpose. A simulation was used to validate accessibility, scalability, security and vulnerability of the proposed authentication framework. Finally, the analysis of results demonstrated that the proposed authentication framework carries the ability to fulfil the requirements of securing the virtual agents in the UK banks from external and internal cyber threats.

Keywords Virtual agents · Ebanking · Authentication · Cyber attack · Google Lighthouse utility · CBEST

1 Introduction

1.1 Aim

The aim of this research is to develop an authentication framework which can be adapted to virtual agents use in the UK banks. To achieve this aim, the researcher

H. H. Hettiarachchige · H. Jahankhani (✉)
Northumbria University, London, UK
e-mail: Hamid.jahankhani@northumbria.ac.uk

245

has explored the domain knowledge around e-banking, virtual agents, authentication frameworks and models to identify the knowledge gap and research questions.

1.2   Background

The use of virtual agents is exponentially growing due to the advancement of information systems and technology. A number of different industries are adapting virtual agents in their regular business operations to increase the efficiency, accuracy and security of services. Additionally, the use of virtual agents helps organisations to reduce the cost of service. The growth of Artificial Intelligence (AI) related technologies has significantly contributed to the enhanced functionalities of virtual agents. According to Soderlund et al. [78], AI-backed virtual agents are capable of performing complex tasks with a number of workflow steps. This emphasis the inevitability of investigating the authentication and security of the virtual agents.

Authentication plays a key role in securing digital applications. Thus, authentication of the virtual agents is required to provide a secure end-to-end digital service. The financial sector is one of the key service sectors which requires a higher level of security, reliability and consistency among the service provided to the customers. A number of UK banks have already adapted virtual agents to serve customer queries and repetitive tasks. However, there is limited research conducted to analyse the authentication of the virtual agents. Kiljan et al. [36], highlighted that there is a room for future research to conduct a systematic analysis of authentication characteristics of different e-banking systems.

An empirical study conducted to analyse the user trust of virtual agents emphasised that security and privacy has a significant impact on consumer trust on virtual agents [26]. According to Nadarzynski et al. [55], a significant proportion of users were not willing to accept AI-driven virtual agents in the healthcare sector due to security and privacy concerns. Moreover, Følstad et al. [26] highlighted future work includes a study of the security and privacy of the virtual agents guided by a theoretical framework.

Dobbie [20] suggests that multiple layers of authentication are required to secure banking systems and applications. Strong Customer Authentication (SCA) is another requirement enforced by the Financial Conduct Authority (FCA) UK to protect banking customers from fraud. Limited researchers have analysed the security requirements of virtual agents form the customers' perspective and the technical or systematic evaluation has not been considered. Therefore, the researcher has identified this as a gap in the knowledge to develop an appropriate authentication framework for virtual agents use in the UK banking sector.

Implementing an authentication framework for virtual agents will provide further opportunities for banking and other financial institutions to generate enhanced capabilities of the virtual agents to provide a better service cost-effectively. Similarly, it will improve the consumer trust to use the virtual agents confidently in their regular banking activities.

2   Literature Review

This chapter critically examines the existing literature on virtual agents, AI and E-banking, security, authentication and authentication models to identify the knowledge gap. Firstly, the chapter starts with the introduction to virtual agents and the significance of virtual agents. Secondly, the significance of E-banking and AI were discussed to provide an overview of the preferred business area. This emphasises the importance of E-Banking in the UK service sector. Thirdly, the literature about authentication, authentication models and security are examined to identify the importance of security and authentication. Finally, the limitations of the authentication methods and models provide a conclusion to findings from the existing literature.

2.1   Virtual Agents

A virtual agent is a computer program or software which can serve customers using artificial intelligence via voice, video or chat functionality [39]. The use of virtual agents as service agents has significantly increased in a number of industries due to the recent advancements in information technology [86]. Trescak and Bogdanovych [85] suggest that the evolution of the technologies also enables virtual agents to act with 'human-like' conversations and perform repetitive tasks on behalf of humans. These systems have the ability to use a number of input sources such as text, voice, image recognition, and contextual information to perform different tasks.

According to Juniper Research [34], it is estimated that there are 3.25 billion digital assistants used worldwide. This is expected to grow up to 8 billion by 2023 [52]. Technology market leaders such as Microsoft, Apple, Amazon and Google also introduced virtual agents such as Cortana, Siri, Alexa and Google Assistant [10]. Virtual agents are widely used in e-banking, health care, tourism, real estate and property, law firms, the tech industry and marketing firms.

A survey discovered that more than 78% of the organisations in the service industries using virtual agents in their operations or considering implementing virtual agents [80]. Rese et al. [70] indicate that the virtual agents in the service industry are also being utilised frequently in activities such as collecting information, providing guidance, recommendations and greeting customers when they call. The author further discovered that the use of virtual agents to assist in simple scenarios is the most popular choice of the service sector [70].

According to Alsop [2], 85% of the IT community respondents from the UK predicted that virtual agents would have significant technology changes in the development of voice, real-time translation and AI-driven chatbots to perform more advanced operations. As an example, Crosman [16], investigated the virtual assistants' capability to help customers to pre-qualify for mortgages by detecting the problems and potential non-compliance. This highlights the potential technology trends in the future and the significance of the virtual agents. However, this level of operations is not widely utilised in the service industry due to several concerns in operations and security [16].

A survey indicates that more than 64% of participants suggest that virtual agents are being used to provide real-time and personalised insights to understand and resolve customer queries [81]. A virtual assistant serves customer queries is inter-connected with different channels, services, API endpoints, devices and knowledge sources [35]. The impact of COVID-19 pandemic has influenced the businesses to re-think their business model to limit face-to-face conversations [77]. This increases the motivation of banks and other financial organisations to utilise more capabilities of virtual agents.

2.2   AI and E-Banking

E-banking provides the facility for the users to perform banking activities from anywhere in the world [27]. It provides a number of advantages over the traditional banking strategy. One of the key advantages of using e-banking services is saving time and cost while providing a fast accessible and reliable service [57]. Therefore, Reddy [67] suggested that rolling out AI-powered virtual agents to facilitate e-banking reduces cost-of-service operations by 30%.

A study conducted to measure the usage of e-banking in the UK highlighted that the average e-banking usage is 64% [27]. According to Garín-Muñoz et al. [27], this is also known as the penetration rate for e-banking. This was conducted using the users between the age group of 16 to 74 years. Besides, it was suggested that the average penetration rate across the European Union (EU) is at 45%. This highlights the significance of virtual agents and e-banking systems in the UK as it is 19% higher than the EU.

Cabrera-Sánchez et al. [11] suggest that technology fear among the consumers affects the relationship between the consumers and service providers. Consumer concerns about privacy and security when using the AI-driven e-banking services and virtual agents emphasise the importance of secure authentication framework. The e-governance helps to maintain the governance regime of the organisations [27].

An analysis of the top 10 biggest UK banking websites demonstrated that only four of them are currently using virtual agents to serve the customers [53]. Furthermore, none of them is using any authentication mechanisms to authenticate virtual agents. According to Kumar et al. [40], the virtual agent of the expected to facilitate a wide range of functionalities. This aims at enhancing the knowledge and experience of the bankers by providing accurate and fast responses to a wide range of queries.

2.3 Security and Authentication

The growth of technology advancements makes challenging for the banks to keep up to date with the security of the e-banking systems. Hassan and Ahmed [30] state that security and privacy is the ultimate concern for any e-banking system. Therefore, virtual agents used in the banks inherits this concern. Moreover, the lack of security models and an increasing number of attacks against financial organisations make virtual agents vulnerable to intrusion and attacks [17].

Dan et al. [17] suggest that adopting a reliable and robust security model is necessary to ensure the information security standards of all the digital transactions. A number of security controls are in place to secure the e-banking transactions. The use of Secure Socket Layer (SSL) is widely used to encrypt the data during the transfer. However, implanting the SSL does not seem to be effective against the fraudsters as the global payment card fraud increased by 19% to reach $14 billion in 2016 [42]. Furthermore, compliance requirements and standards create additional pressure against the financial organisations to consistently evolve with the security of the online systems.

Authentication is the process of verifying the legitimacy of the user by validating the user details in different forms such as username, password, biometrics, voice, video or photograph or a security question [41]. Moreover, the authentication provides the ability to trust the user or customer. However, authentication does not provide the details on what the user can do or what documents, or transactions they can access.

Biometrics authentication is used in several UK banks such to authenticate users using mobile apps [20]. According to Dobbie [20], authentication using the biometrics has proven to be not 100% secure. Ethical hackers at the GeekPwn 2019 conference demonstrated that it is possible to compromise any smartphones fingerprint scanner in under 30 min. Therefore, it is suggested that additional authentication layers are required to secure the banking customers to enhance the security level [20].

Königstorfer and Thalmann [42] suggest that AI is closely related to machine learning and data mining. Machine learning can be defined as the ability of a computer system to adapt and learn based on the patterns of data. The data mining is the analysis and understanding of data to discover important information [42]. Increased behavioural changes of fraudsters and number of security breaches influence the banks to improve the security and utilise AI capabilities to the e-banking platforms.

The deepfake fraud is another form of challenge that banks and other financial organisations are facing in when authentication of the users [93]. Deepfake is an AI-based technology which can produce fake content or video [84]. Moreover, the practical scenarios of deepfake involve in the manipulation of legit content to trick the authentication systems which uses videos, voice or images.

Guo et al. [28] state that researchers demonstrated the failures to preserve integrity in most of the mobile e-banking mobile applications. The Confidentiality, Integrity and Availability (CIA) are the foundation factors of information security. Failures to secure integrity means that e-banking applications and services can be vulnerable to unauthorised modifications [28].

Technological factors such as convenience, the authenticity of the conversation and risk factors such as privacy concerns and immature technology have a positive effect on behavioural intention [70]. The study analysed the behavioural intentions for the chatbot 'Emma' demonstrated that virtual agent users are concerned about the privacy, security and authenticity of the conversations of virtual agents. Tuza [87], suggests that there are limited academic papers conducted to study the security of the virtual agents against data theft, Man-In-the-Middle (MITM) attacks, data leakage and potential data breaches.

The authenticity is defined as the ability of a virtual agent to have a natural conversation. There is a possibility that such a conversation may exchange sensitive data and personally identifiable information. Therefore, the virtual agent needs to be protected with an appropriate authentication framework. According to Rese et al. [70], an analysis conducted using modified TAM (Technology Acceptance Model) and Uses and Gratifications (U&G) theory discovered that authenticity of the conversation with the virtual agent has a significant positive effect on the intention to use the virtual agent.

2.4 Authentication Models

OpenID is a framework which allows users to authenticate with relying on parties using third party identity vendors [58]. For instance, if a user needs to book a hotel from a booking site, and the booking site and supports OpenID connect, the user has the ability to use login with identity providers such as Google, Facebook, Twitter and Amazon [56]. The OpenID has the ability to add an authentication layer using OAUTH with identity details. OAuth is a framework which can be used for authorisation with RFC6749 standard. According to Dodanduwa and Kaluthanthri [21], OpenID connects it mandates the involvement of a human being in the authentication process. However, there was no satisfactory evidence found on adapting the OpenID Connect into a framework which can authenticate the virtual agents.

There is a number of access control models, methods, technologies and administrative capabilities used to design the access control systems [96]. However, every access control system is unique depending on the business context and security requirements. According to Younis et al. [96], the National Institute of Standards and Technology (NIST) suggests that any access control system should have the capability, privileges, object, subject and action as elements of the system. The differences in commercial security policies contribute to the discovery of different access control models. Furthermore, the Mandatory Access Control (MAC) and Discretionary Access Control (DAC) are two different access control models developed based on different requirements.

A number of flows have been discovered in the MAC and DAC access control models [15]. The MAC model is administrator centric and it carries the limitations of maintainability, scalability and user-friendliness [72]. Therefore, it is not suitable for adept as an authentication framework to secure the virtual agents used in the banks. However, Conrad et al. [15], suggests that the MAC model still can be used in the scenarios to handle the classified government data.

Rivera Sánchez et al. [72] suggest that the DAC model provides a certain level of control over the data for users. Access Control Lists (ACLs') are used to define the access permissions to control the data for specific user groups. Therefore, this model may contain a low level of data protection as the users with access has the ability to share the data as they like. Furthermore, the obscure nature decentralises the access management which can lead to a nightmare of managing the ACLs in complex use cases [15].

The Role-Based Access Control (RBAC) model has certain advantages over the MAC and DAC models. The RBAC model is based on the pyramid approach with a set of defined roles and their respective access levels (Rivera Sánchez et al. 2019). Furthermore, the RBAC model is recognised as the best practice model for authenticating and managing scenarios such as operating systems and other software.

Aliane and Adda [1] state that attribute-Based Access Control (ABAC) model is one of the most popular access control models over the past decade. ABAC model manages the access using attributes of each system components. This enables the ABAC model to be integrated with complex business systems [1]. Therefore, ABAC model can be deployed across many financial organisations such as banks and insurance companies.

The Risk-Based Access Control (R-BAC) is based on the risk factor of a certain access request decision [3]. Therefore, a risk analysis is performed to evaluate each access authentication request. Atlam and Wills [3] argue that the estimated risk factor is then compared with the access policies to determine the access decision. However, risk estimation is not a straightforward task. Factors such as type of access, location of the request, data sensitivity requested, and the access history of the user need to be considered to construct the decision. Therefore, it can be difficult to adept as it is for the authentication of the e-banking or virtual agents.

The Rule-Based Access Control (RAC) model can be useful in the scenarios where the access can be controlled based on the standard set of rules such as filtering based on whitelist or blacklist (Rivera Sánchez et al., 2019). This can be suitable for more simple and straightforward access request scenarios such as whitelisting access to a computer-based on the IP address or block requests from a particular IP address.

A number of access control models are used to secure internet banking applications [69]. Authentication, Authorisation, Accounting framework (also known as the Triple-A framework) is an intelligent framework used for controlling the access to computer systems by setting the policies, providing the usage audits and information.

The generic AAA server architecture consists of AAA server or service which, application-specific modules, policy and events repository [66]. According to Kolluru et al. [37], triple-A model itself is not sufficient enough for the systems which interact with other services as the security could be improved by adopting a mutual authentication between the different services along with the triple-A model. This model is based on extendable and open architecture and allows to use Extensible Authentication Protocol (EAP).

2.5 Authentication Methods Use in Online Banking

It is important to discuss the authentication methods used in the online banking systems in order to understand the authentication framework required for virtual agents.

2.5.1 Username and Password

Username and Password is the most commonly used authentication method to authenticate users in digital applications and devices over the world [32]. Typically, the user has the ability to register themselves with a username and password prior authenticating into the applications or services. Therefore, username-password authentication requires a memorised password from a client.

However, username and password authentication can be vulnerable to password guessing attacks and brute force attacks [32]. Furthermore, Bani-Hani et al. [5], highlighted that cyber-attacks against banks motivated most of the banks to look for stronger authentication methods rather than normal username and password. Therefore, username and password authentication itself will not be sufficient against the authentication of virtual agents as well.

According to Bani-Hani et al. [5], the authentication using the username and passwords can be strengthened by implementing appropriate password policies and Multi-Factor Authentication (MFA). The verification PINs are frequently used in ATM transactions and e-banking systems and they still can be vulnerable against keyloggers and keystrokes [5].

2.5.2 API Keys

API keys are being used in the banking systems to authenticate different endpoints in the backend application services [14]. These keys are being used programmatically with no end-user involvement. Chubarov et al. [14] suggest that a banking system can use a picture of the customer and then it can be passed to a separate subsystem for facial recognition. The communication between the main banking system and facial recognition subsystem can utilise the API keys for authentication. However, API keys are not suitable to use in end-user authentication as the intruders may be able to extract them from the browser.

2.5.3 Token-Based Authentication

According to Haekal and Eliyani [29], token-based authentication allows users to use their credentials to obtain a token from the system. The token obtained then can be used to access a particular resource. Handling the token-based authentication in multi-platform scenarios might be complicated with the server-based authentication [29].

Indu et al. [33] suggest that additional encryption needs to be incorporated to secure the token-based authentication. Furthermore, the Security Assertion Markup Language (SAML) can be used to improve the token-based authentication. This highlight additional factors need to be considered when adapting token-based authentication for secure applications. Therefore, token-based authentication needs to be carefully considered before adapting to the authentication of virtual agents.

2.5.4 Biometrics Authentication

Biometrics authentication is popular in the service industries where it requires a higher level of security [94]. According to Bian et al. [8], fingerprint authentication the mostly used biometrics authentication method. Modern smartphones have fingerprint scanners in place to perform the authentication based on fingerprint scan [8].

Facial recognition is another popular authentication mechanism used in financial banks [4]. According to Lupu et al. [47], facial recognition data is recorded in most of the UK banks during the process of opening a bank account. This data is then used in a number of occasions where the bank needs to verify the user identity.

Besides, voice recognition is also can be used to verify the identity to facilitate the user authentication process. Some banks currently using voice recognition capabilities in their operation to verify the identity during the user authentication process [13]. However, a survey conducted by banks revealed that only 18% of respondents trusted voice and facial recognition [13]. Therefore, using facial or voice recognition as the primary identity verification or authentication mechanism can be questionable.

Banking applications should contain the ability to leverage the capabilities of biometrics authentication provided by smart devices. Based on a survey conducted, it was discovered that 92% of the UK banking consumers and 69% of US consumers prefer the biometrics technologies for identity verification and authentication.

Therefore, adapting biometrics capabilities to authenticate virtual agents can help to maintain the common interest of authentication between different systems. However, it was highlighted that users have privacy concerns about using biometrics authentication. An evaluation of privacy concerns related to the biometrics data collection, handling, secondary usage, errors and unauthorised access will be required to satisfy the concerns of the users.

2.6 Authentication in Cloud-Based Applications

The challenges on cloud-hosted applications need to be evaluated when considering an authentication framework for the virtual agents [89]. According to Vassilev et al. [89], the majority of the UK banks who uses virtual agents also uses public cloud infrastructure to host the applications. The data confidentiality, security and privacy, authentication and authorisation, the security of the infrastructure and network, identity management and single-sign-on has been identified as the challenging areas which require further research [22].

Furthermore, Ethelbert et al. [22], suggest that many researchers are facing the challenges of account hijacking, verification and protection of credentials and data breaches on cloud-based applications. Moreover, it is important to secure the communication and transactions processed through virtual agents by adapting an appropriate authentication model. The security of the cloud application is a much wider area and beyond the scope of this research.

2.7 Open Banking

The strict controls in the UK banking industry and heavy regulation requirement demands the ability of the banking data to be securely shared via Application Programming Interfaces (API) with third parties [63]. Therefore, the concept of Open Banking was introduced by the Competition and Market Authority (CMA) to address this requirement across the UK banks. The use of API's enables the banks to communicate with different core systems to perform the tasks required. As per the European regulative standard, the Open Banking API must follow the Payment Service Directive (PSD2) standard to provide access to third-party providers [64].

According to Premchand and Choudhry [64], Open Banking framework consists of four key building blocks named as APIs, Data, Operation Governance and Security. The APIs contains the definitions of the interfaces, development, testing and maintenance of the API engines. The data includes the definition, structure, permission, access rights and the format required for the sharing. Operation and governance building blocks consist of monitoring and administration lifecycle management, how the customer complaints are handled and how the data is secured once shared. The security block covers the access level permissions and encryption including authentication and authorisation.

The authentication framework for the virtual agents should contain the ability to follow the requirements of Open Banking framework to facilitate the requirements in the building blocks. Furthermore, this also highlights the requirement of an authentication framework for virtual agents, as they should comply with the PSD2 standards when performing account-related activities against the banking customers. This highlights the importance of Open Banking framework when creating an authentication framework for the virtual agents used in the banking industry.

2.8   Limitations of Existing Authentication Models and Methods

The critical analysis of existing access control frameworks and authentication models demonstrated that the current authentication models do not have the capability to service the authentication of virtual agents by out-of-the-box. This is because the authentication models itself does not provide the required capability, security, privacy and extendibility to integrate with the virtual agents. Furthermore, current models do not consider the attack vectors and vulnerabilities that e-banking systems are experiencing.

According to Bani-Hani et al. [5], online authentication methods used in banks are consistently getting attacked against these methods. A research conducted to examine the authentication methods and the attacks against them in UAE identified 16 common attack types [5]. These attacks include social engineering attacks, keyloggers, shoulder surfing, screen capturing, dictionary attack, brute force attack, phishing attack, hardware observation attack, MITB attack, network sniffing and MiTM attack. Therefore, it is important to develop an authentication framework which has the ability to adapt and facilitate the mitigations against the attacks (Table 1).

In summary, the critical literature analysis of the virtual agents use in the banking industry emphasises the requirement of an authentication framework to secure the transactions involving virtual agents. The analysis of existing authentication frameworks and models demonstrated the absence of an adaptable authentication framework for virtual agents. As elaborated in the literature review, anonymous virtual agents can be vulnerable against a number of cyber-attacks. Therefore, the proposed framework should have the ability to provide federation between a number of other e-banking systems while securing the authentication, security and privacy of the virtual agents.

Table 1 Comparison of authentication models

	Beacon	PWD	Token	Bio	Behav
Easy to use	✓		✗		✓
Add device		✓	✗	✗	✓
Not sensitive	✓		✓	✗	✗
High security	✓	✓	✓		
Reliability		✓		✗	✗
Speed			✓		✗
Cost		✓	✗	✗	
Trending	✓			✓	

Source Vassilev et al. [89]

3 Methodology

3.1 Research Process

According to Saunders et al. [74], the most common research approaches are deductive and inductive approaches. The deductive research approach is usually based on quantitative analysis and inductive research approach follows a qualitative analysis [74]. However, in certain scenarios, both approaches can be applied depending on the researchers' vision or the perspective of the study. Moreover, Saunders et al. [74] state that this approach is known as a sequential mixed method. Sequential mixed methods comprise of multiple phases of data collection and analysis.

In this research, the development of a theoretical framework follows an inductive research approach where the researcher uses knowledge gathered from the existing literature to develop an untested theory or a framework. Then, the framework validation uses a deductive approach with quantitative data extracted from a simulation package. Also, the researcher will use a certain method followed by another to expand and justify the findings. Therefore, this research follows a sequential mixed methods research design (Fig. 1).

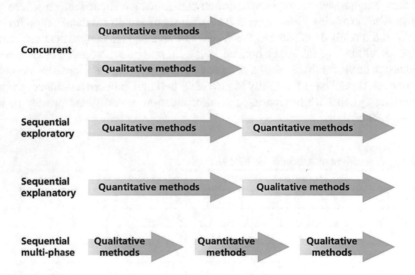

Fig. 1 Different types of mixed methods. *Source* Saunders et al. [74]

3.2   Design of the Authentication Framework and Experiments

The proposed framework is designed with a high-level diagram followed by an authentication sequence diagram to elaborate on the process of authentication and authorisation of virtual agents in the UK banks. The new framework consists of authenticators, capabilities of the virtual agents and other e-banking systems that require interaction with the virtual agents. In addition, it employs the authentication best practices identified by prior researchers conducted in different industries.

Moreover, designing an authentication framework for a UK banking industry requires a number of considerations related to the CIA. The researcher considered six core elements when designing the framework to ensure the CIA.

Types of credentials is the first element considered when designing the proposed framework. This represents what type of credentials can be used with the proposed authentication framework. Management of credentials is the second element which denotes how the credentials are going to be managed in the suggested framework. Monitoring of the authentication process is required to ensure the accountability of the proposed framework. Next, authentication service characteristics are considered to ensure the integration capability and authentication with other e-banking applications. Finally, information governance desires are considered to ensure the governance requirements are met.

Additionally, the proposed framework also considers the guidelines from PSD2, open banking, FCA, CBEST and DPA to comply with the required standards of UK government and financial industry. A detailed discussion of how the research has practically occupied above standards and core elements for the proposed framework can be found in chapter "Socio-technical Security: User Behaviour, Profiling and Modelling and Privacy by Design".

Experiments are designed to analyse the expected behaviours and practicality of theories and frameworks. In this study, the validation of the proposed framework involves creating an experiment which contains a systematic review with quantitative analysis.

The researcher has formed a simulation environment which is closer to a real-life scenario to run the steps of the proposed framework. A virtual agent is configured and hosted on researchers personal Amazon Web Services (AWS) cloud account. The author has integrated the virtual agent with the Amazon Cognito which is a scalable identity provider. The concurrent tests are designed and developed with samples sizes of 100, 250 and 500 requests to analyse the authentication behaviour.

3.3   Implementation of the Simulation Environment

A simulation was implemented to extract the data required for this research to validate the proposed authentication framework. The simulation represents an environment with core authentication functionalities of the proposed framework. Also, this can be identified as a Minimum Viable Product (MVP). According to Nuseibah et al. [59], MVP can be used to represent a complex and challenging environment.

Figure 2 illustrates the environment created in the researchers personal AWS account in US-East-2 region. This environment consists of one EC2 instance with 8 CPUs and 16 GB RAM, Amazon Cognito user pool and application load balancer. The Bot Framework Emulator, Apache JMeter and OWAZP ZAP tools are installed and configured in the EC2 instance. The Bot Framework Emulator service is exposed via an Application Load Balancer (ALB) as an endpoint to conduct the JMeter tests and security scanning tests. The Bot Framework Emulator is integrated with Amazon Cognito to provide the capabilities of MFA and OAUTH 2 to the Bot Framework Emulator simulation.

Apache JMeter tests are configured using a static token as MFA forbids the use of systematic API calls to perform the authentication requests. However, this does not affect the actual ability of authentication using MFA. The OWASP ZAP vulnerability scan uses the open-source version of 2.10.0 to perform the vulnerability scans against the load balancer endpoint. Appendix E contains the further configuration details of the simulation environment.

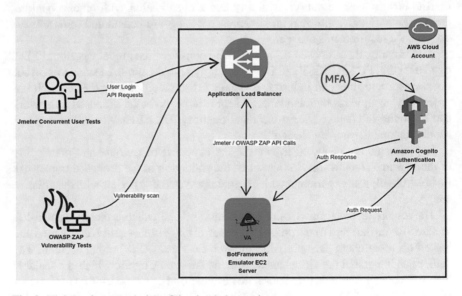

Fig. 2 High-level system design of the simulation environment

3.4   Definition of Validity Variables

Validation variables are defined to evaluate the effectiveness of the proposed virtual agent authentication framework from a technical perspective. These variables are identified as accessibility, scalability, security and vulnerability.

3.4.1   Accessibility

Accessibility is the ability to access the software or utility by a wider range of people. According to Kiljan et al. [36], accessibility contains aspects of convenience and inclusivity. The convenience includes the authentication time and the inclusivity contain additional support for disabled groups. Accessibility is measured using the Google Lighthouse reports. This uses the Web Content Accessibility Guidelines (WCAG) 2.0 to analyse the accessibility criteria of the simulated environment.

The accessibility crucial for the functioning of this authentication framework as the end-users should have the ability to access virtual agents using user devices such as mobile phones. According to Wille et al. [92], accessibility metrics provides different benefits such as the ability to measure the accessibility level of components, providing the quality assurance and ranking of the web pages.

3.4.2   Scalability

Scalability of an authentication framework is the ability to scale up to serve a large number of concurrent requests [90]. Also, Barzu et al. [7] suggest that scalability is the ability of an application to scale up horizontally and vertically. This emphasis that the authentication framework should have the ability to serve a growing number of concurrent users by facilitating with application and infrastructure scaling. Furthermore, it is also depending on the scaling time of infrastructure as the infrastructure should support the scalability of the application for smoother user experience [7].

The ability to scale becomes extremely crucial when no human interaction is involved in the process. It is important to understand the impact of the login process from a quantitative perspective to analyse the potential end-user experience after the authentication framework is adapted. Therefore, the authentication process and token handling process need to be solid and scalable.

Scalability test scenarios were decided after considering the available physical resource in the simulation environment. The test samples were selected based on the simulation server specifications conducted to identify the acceptable physical resource utilisation. Therefore, samples of 100, 250 and 500 concurrent user tests are conducted to examine the ability to serve without overloading the system resources.

3.4.3 Security

It is important to understand the security to take preventive and mitigation actions against cyber-attacks related to authentication. In this research, the security is considered from the perspective of predictability, abundance and disclosure of the authentication methods of the proposed framework. Security is evaluated using qualitative analysis of security properties and respective controls implemented. This provides an overview of security implementation configurations required to consider when implementing the proposed authentication framework.

3.4.4 Vulnerability

Vulnerability covers the confidentiality and privacy aspects of the proposed framework. This is measured using the number of vulnerabilities detected from the vulnerability scans. Analysis of vulnerabilities in the proposed framework was conducted using OWASP ZAP utility. According to Makino and Klyuev [48], OWASP ZAP is used widely in the industry for vulnerability testing of service-oriented applications. Therefore, the tool is used to run vulnerability tests against the simulation environment.

The OWASP ZAP tests are conducted in the environment to check for any potential vulnerabilities in the proposed framework. The test reports classify the vulnerabilities detected based on the severity (High, Medium, Low). In certain scenarios, the vulnerabilities detected can be falsely positive or not applicable for the particular scenario. Therefore, a manual review of findings is required to verify the detection data.

3.5 Tools Required to Validate the Proposed Authentication Framework

This research uses both open source and proprietary tools to design and validate the framework. The virtual agent was configured using an open-source virtual agent software known as Bot Framework Emulator. The Bot Framework Emulator software does not provide the ability to use MFA authentication by out of the box. Therefore, the researcher has integrated Bot Framework Emulator application with AWS Cognito authentication framework. The AWS Cognito authentication service is a proprietary service developed by Amazon Web Services to provide the ability to integrate with OAuth 2 providers.

Google Lighthouse utility (V6.0) was used to detect the accessibility criteria of the proposed authentication framework. The accessibility tests are conducted by the audit tests executed against the API endpoint of the virtual agent. The lighthouse utility provides an accessibility report including the overall rating and number of accessibility issues with their respective details.

Apache JMeter is an open-source tool used to run the concurrent test scenarios against the simulation to perform authentication requests. This tool is widely used in the industry and the researchers to simulate concurrent requests against web services or applications [82]. In this research, JMeter executions are conducted against the Bot Framework Emulator API endpoint by triggering the authentication requests to evaluate the scalability of the proposed framework.

According to Kritikos et al. [38], OWASP ZAP scanner is a widely used utility by the industry and the researchers for identifying the vulnerabilities in web-based applications. The scans are pointed to the API endpoint of the Bot Framework Emulator virtual agent. As the key focus is on validating the proposed authentication framework, security is discussed in a holistic perspective rather than analysing the different attacks and suggesting the mitigation actions.

3.6 Validity and Reliability of the Test Variables and Tools

Saunders et al. [74] argue that the value of research findings depends on the reliability and validity of data collection technique or analysis procedure of research. Moreover, the authenticity of the proposed framework depends on the suitability of the variables that use to measure it.

The test variables identified in this research are also used to evaluate authentication frameworks in different industries [9, 36, 51, 68]. In addition, some of the previous researchers have suggested a systematic analysis of these variables can be conducted in future research work [36].

The tools employed to extract the data from the simulation are industry standard tools which are being used by other researchers and organisation as well. Lenka et al. [43], highlighted that Apache JMeter can be used for performance analysis of web services and applications. JMeter can also be used to evaluate the scalability of an application by conducting stress tests against a web-based application [61].

According to Devi and Kumar [19], OWASP ZAP can be used to test the vulnerabilities of web applications. It was highlighted that these tools are used to evaluate the vulnerabilities in e-banking applications. Furthermore, OWASP ZAP is widely used by the researchers and industry experts to perform security and vulnerability analysis of web applications. Therefore, it can conclude that the variables used in this research and the analysis techniques provide a realistic representation of data extracted.

3.7 Legal and Ethical Considerations

As this research uses a mixed methodology, the data collected from the existing researchers follow the ethical guidelines of the British Research Association (BERA). Therefore, the secondary data is not amended, updated or modified by the researcher. The data collection was conducted using the open-source Bot Framework Emulator simulation package. The simulation uses dummy data; hence the environment does not contain any sensitive or personally identifiable information.

The Amazon Cognito authentication service is a proprietary authentication engine developed by Amazon and available for evaluation use in the AWS Cloud services platform. Therefore, the researchers' personal AWS cloud services account and the personal laptop are used to configure the integration of authentication with Bot Framework Emulator. The authentication tests are conducted using random usernames and passwords generated programmatically. Therefore, data collection does not have any risks of compromising the actual user details.

The details and the names of the banks mentioned in this research are obtained from the previous researchers and publicly available information via the respective institutions' websites, FAQ pages, and manuals. Therefore, it does not reveal any sensitive data related to the banks in the UK.

The ethical considerations of using a simulation to represent a practical real-life authentication scenario do not cover all the aspects of the authentication requirement. However, the scope of this research is limited to the development of the authentication framework and validation of the proposed framework within the boundary. Therefore, the researcher has limited the number of attributes associated with the suggested variables. The data required for the attributes used in this research not contains any adverse effect of using a simulation to collect the data.

In conclusion, a mixed-method was used as the choice of methodology for this research project. The researcher used a qualitative analysis to develop the proposed authentication framework. Afterwards, a quantitative analysis was conducted to validate the proposed framework. The data collection was conducted by performing authentication tests against the simulation environment to extract the data for validation variables of accessibility, scalability, security and vulnerability. Google Lighthouse, Bot Framework Emulator, Apache JMeter, OWASP ZAP and AWS Cognito tools were used to facilitate the simulation of the study. This chapter ends with the discussion of legal and ethical considerations of the study.

4   Data Analysis and Critical Discussion

4.1   Analysis of Existing Virtual Agents in the Major UK Banks

The critical discussion starts with an analysis of virtual agents in the UK banks to investigate the importance of having an authentication framework. This analysis ultimately provides the answer to the first research question of the study. During the analysis, the researcher discovered that only a few banks have already implemented virtual assistants and the rest of the banks are still in the planning stage.

Several well-known banks are currently employing the virtual agents in their online banking services. These virtual agents are utilised for standard repetitive operations such as providing generic guidance, helping the customers with navigation, locating the nearest branch, providing the answers for FAQs [54]. The researcher has selected few of these banks to conduct an investigative analysis of the virtual agents used. This was conducted using publicly available data from the respective banking websites, help pages and manuals.

Virtual agents in some banks use to serve the customer requests anonymously without any authentication mechanism. On the other hand, the virtual agents use in other banks alert customers not to submit any sensitive data to virtual agents. However, there is no appropriate authentication mechanism observed in these banks as well. In case of any request requires authentication, they are being forwarded to a human agent.

The above investigation demonstrates that access to UK banking virtual agents consists of a number of concerns related to security, privacy, legal and ethical areas due to the lack of authentication mechanism. According to Leung et al. [45], virtual agents have an increased demand to provide a personalised and enhanced experience for end-users. As discussed in the literature review, the absence of an authentication mechanism impacts the capabilities of the virtual agents. Thus, leaving them to be served with generic responses or forward the requests to a human agent. This emphasis that, having an authentication framework is vital to fully utilise the virtual agents effectively and securely in the banks.

The risk of exposing sensitive data is a well-known security concern related to the systems with weak or no authentication. Puthal et al. [65] suggest that these systems are vulnerable against a number of attacks. Therefore, having a strong authentication framework is vital to provide defensive mechanisms against these attacks to reduce the risk of exposing sensitive data.

The legal obligations which affect financial institutions have strict conditions against the security and protection of customer data. In addition to the Data Protection Act of 2018 (DPA), financial institutions in the UK have additional governing bodies such as The Financial Conduct Authority (FCA) and Prudential Regulation Authority (PRA). Schydlowsky [75] argues that these regulations consist of additional policies such as the declaration of secrecy, record management and classification, ethics and misconduct policies. According to FCA UK [24], all the UK banks expected to support strong customer authentication for online banking. Therefore, the virtual agents communicating with online banking applications should also facilitate this policy.

4.2 The Proposed Authentication Framework

Figure 3 depicts a holistic authentication framework for virtual agents in the UK banking industry. The proposed holistic authentication framework consists of several key components which are customer, virtual agent, authenticators, monitoring, information governance, bank and e-banking services. In addition, the validation variables are highlighted with green colour.

The authentication process is initiated based on a service access request from a user. For instance, this can be an enquiry about a transaction, payment information or any other information that virtual agent has to obtain from different subsystems within the domain of the bank. Once the customer request is sent, the virtual agent forwards that to the authentication. The authenticators then perform the authentication and issue a security token for the request.

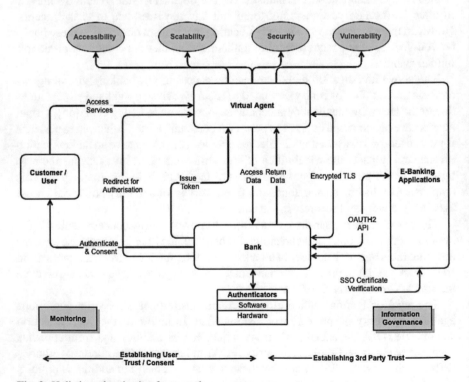

Fig. 3 Holistic authentication framework

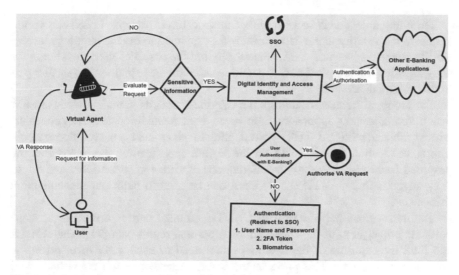

Fig. 4 Authentication process of the standalone subsystem

The proposed authentication framework was designed as a standalone subsystem (Fig. 4) which can be provisioned and integrated with other e-banking systems as required. This provides the ability to reuse the framework across several different banks and banking systems in the UK. The proposed framework mandates the user login and validation for any requests or responses which can contain the sensitive data. Therefore, it can ensure the privacy concerns of the conversations with the virtual agent.

Figure 4 demonstrate the standalone authentication process of the proposed framework. The authentication process consists of multi-factor authentication steps. The MFA authentication can be conducted via hardware or software authenticators. Furthermore, it has the ability to plug the biometrics authentication methods as required to leverage the biometrics authentication in the user devices.

According to Bani-Hani et al. [5], a study conducted to evaluate the authentication methods in the banks highlighted that use of software and hardware tokens are the current most popular choice to authenticate the e-banking applications in the UK. The biometrics are used to provide the access to the token generation device or application. Therefore, authentication of the virtual agents can also leverage this to reduce any additional work required to roll out the proposed framework.

An authenticated virtual agent can provide advance capabilities such as responding to the transaction-related enquiries, performing checks and validations against the account and payment related tasks. Popular virtual agents such as Amazon Alexa, Google Assistant and Siri has the ability to perform financial transactions on behalf of the users [50]. However, this level of advanced capabilities is not yet fully utilised in the virtual agents of the UK banking industry.

The communication is secured with Transport Layer Security (TLS) encryption to ensure the security in the data transfer. As explained in the methodology, accessibility, scalability, security and vulnerability are the variables used to validate the authentication framework. The data for these variables are explained in the latter part of this critical discussion.

The proposed framework follows the OAuth2 standards to maintain consistency over different banking applications. However, there are additional security considerations evaluated with OAUTH2 when adapting to the virtual agents for this research. The authentication flow is based on the OAuth specification and adapted to the proposed framework to secure the interactions of virtual agents. Richer and Sanso [71], suggests that OAUTH specification can be used to build our authentication framework.

The transactions performed with OAUTH protocol carries a number of steps before the delegation of the authorisation grant and return data [71]. Therefore, it can limit the exposure of the data as per the need of each party involved in the authentication process. This provides the ability to further secure the virtual agent transactions in the banking environment.

4.3 Functionality of the Proposed Framework

Figure 5 illustrates the authentication and authorisation flow sequence of the proposed framework.

The authorisation flow consists of four key roles identified as resource owner (virtual agent), client, authorisation server and the banking application server. Virtual agent (resource owner) is the role who controls the resources being accessed by the customer. Banking application server (resource server) responsible for protecting the resources. The client represents the user who wishes to access banking resources. Access tokens are issued by the authorisation server for the customers upon successful authentication. According to Richer and Sanso [71], permission grant follows the authorisation scope defined in the authorisation server. Therefore, this needs to be pre-configured to limit access to the corresponding activities of the specific user account.

One of the advantages of the proposed authentication framework is allowing customer consent during runtime. This is achieved by building a new authentication framework on top of OAuth authorisation specifications. According to Sendor et al. [76], this provides the ability for the user to decide the applications or services they need to access securely. Furthermore, it also carries the ability to scale across multiple different security domains.

Access to different e-banking APIs can be achieved using delegate access capabilities provided by the framework. This enables the integration of existing API is driven banking services with the authentication framework. The banking applications and application servers get the benefit of being able to validate the customer

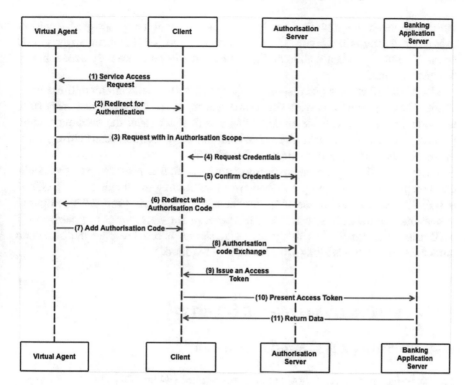

Fig. 5 Authentication/Authorisation flow sequence

requests before sending the responses. This ensures the confidentiality and integrity of the request handling process of virtual agents in the banks.

The proposed framework relies on the MFA to generate access tokens to perform the authentication. In the proposed framework, the authentication and authorisation occur in a separate subsystem. Therefore, authentication and the authorisation process are coordinated by the virtual agent. This creates an opportunity to have a hijack the sessions or tokens with the Man-In-The-Middle (MITM) attacks. Prapty et al. [62] argue that sessions can be protected using one-time encrypted session cookies. Therefore, this framework uses encrypted constrained TLS tokens to secure token authentication. This approach has been identified by a number of previous researchers [25, 79, 88]. The tokens are issued as one-time tokens for a limited validity period to minimise the risk of token compromise.

Exposure of the authentication API endpoints requires protection, and they should have a limited exposure only for required resources. The use of Access Control Lists (ACL) can be utilised to archive the limited exposure to the particular resources required. According to He et al. [31], limiting the exposure of the APIs and monitoring of vulnerabilities can help to secure the endpoints. This provides the ability to minimise the attack surface against the authentication and authorisation API endpoints. The API endpoints also need to be monitored regularly for any potential vulnerabilities. The simulation configured to demonstrate the proposed framework uses security groups to whitelist the API access from the load balancers to the virtual agent. Therefore, authentication and authorisation API calls are only accessible by the virtual agent.

Use of signed request and response objects assures the authenticity of the conversation. The proposed framework suggests that every user request and responses from the virtual agents need to be signed using certificates to seal the communication. Several prior researchers have highlighted that singing the requests can help to protect the authenticity of the communication [60, 73, 95].

Adaption of the appropriate MFA mechanism is vital to provide strong authentication capabilities to the proposed framework. According to McDowell [49], GDPR and PSD2 recommend the use of MFA in authentication frameworks. The proposed framework has the ability to leverage hardware or software devices to perform the MFA using one-time token assigned. Furthermore, it can leverage the biometric authentication methods along with the MFA if required.

4.4   Evaluation of the Proposed Framework

4.4.1   Guidelines of the Proposed Framework

The proposed framework follows good practice guide 44 and 45 (GPG 44 and 45) of the UK government as the fundamental building blocks. The GPG 44 is also known as the guidance of how to use authenticators to protect an online service. The GPG 45 provides guidance on verifying and validating the identity of an individual. Identity verification carries the process of confirming the user's identity using different types of information [12]. The researcher has adopted the GPG 44 and 45 guidelines in the proposed framework through six core elements (Fig. 6).

Types of credentials consist of selecting the most suitable and practical credential for virtual agents. The selection of credentials depends on the security level required for the virtual agents and preferred choice for this research is based on MFA authentication.

The management of credentials requires an adaptation of a suitable password policy to ensure administration of the credentials. The password policy addresses the concerns such as length, strength, retention, age, complexity requirement and lockout policy of the credentials. During the literature review, it was highlighted that username and passwords are vulnerable against a number of attacks. However, adopting a suitable password policy can help reduce the risk of cyber-attacks.

As illustrated in the proposed framework diagram, monitoring of the authentication process is required to ensure the quality of service provided to the customers. Monitoring is a vital role in the entire infrastructure, and it can provide insights about the behaviour of the authentication process. This can include factors such as the number of logins, number of invalid logins attempts and last successful login

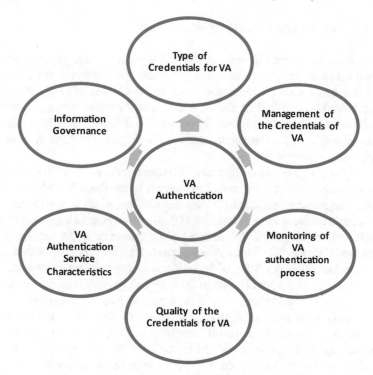

Fig. 6 Elements of the virtual agents' authentication framework

date. Authentication infrastructure requires to be monitored using appropriate monitoring tools or utilities to ensure that there are no bottlenecks exists. Furthermore, monitoring is required to maintain accountability due to legislative and governance requirements.

The quality of credentials in the proposed framework can be ensured by taking suitable measures to prevent from compromisation. These measures require rules to avoid prediction and duplication to mitigate tampering of the credentials. Therefore, this research recommends cryptographic modules to maintain the quality of the credentials. As explained in the types of credentials, adapting an appropriate password policy can also help to improve the quality of the credentials.

Authentication service characteristics help to understand the fitness of the authentication framework in the banking context. The authentication framework suggested for virtual agents should support federation across different e-banking systems as virtual agents will be interacting with different subsystems to perform certain tasks. This was achieved with the use of OAuth 2.0 authorisation protocol support.

The information governance criteria for financial institutions in the UK contains certain regulatory and compliance requirements. The proposed framework has the ability to evolve and facilitate information governance. This is future explained in the compliance of the proposed framework section.

4.4.2   Compliance of the Proposed Framework

Proposing an authentication framework for a UK bank or financial institution requires consideration of several compliance requirements. This involves compliance and legal requirements from the authorities such as FCA, Bank of England—CBEST, PSD2 open banking and GDPR. The FCA insists on the use of strong authentication mechanism and the use of certificates to encrypt the transactions [23]. As discussed in the functionality of the framework section, the proposed framework is designed to satisfy these requirements.

Adapting a Cyber Threat Intelligence (CTI) framework can help to detect and identify any attack against the authentication systems in the financial industry. Wagner et al. [91], suggests that adapting a CTI framework is necessary to get the threat intelligence to build secure systems. The Bank of England has created a framework known as CBEST which is based on CTI to build and improve the cyber resiliency of the financial institutions in the UK [6]. Adapting a suitable CTI framework is suggested during the actual implementation of the proposed authentication framework.

The proposed framework consists of security and vulnerability aspects added to the virtual agent to provide resiliency against the cyber-attacks and gain the vulnerability insights using automated security scanning services. This is further elaborated in the validation of the proposed framework section. Open Banking and PSD2 compliance require the use of secure APIs and secure transactions [64]. Due to the complexity the cost, the simulation does not contain real-time cyber-attack mitigation configurations. However, in a practical scenario, an appropriate Web Application Firewall (WAF) can be adapted to further secure the authentication framework.

The proposed authentication is exposed via restricted and secure API endpoints with TLS encrypted transactions to provide the compliance required for PSD2 and Open Banking. The impact of GDPR is applicable when handling personal data. The proposed framework encryption at rest and encryption during data transfer to ensure the CIA.

4.5   Analysis of Results of the Experiments

Validation of the framework provides quality assurance to the proposed authentication framework. According to Leung [44], examining the validity, reliability and generalisability is important to maintain the eminence of a study.

As described in the methodology, the validation uses the data extracted from the simulation configured by the researcher. The validation tests are conducted locally, within the Amazon EC2 instance. Therefore, the tests do not contain any additional network latency or bandwidth bottleneck. In this research, accessibility is evaluated using Google lighthouse reports and scalability variable is evaluated using performance tests of 100, 250 and 500 user samples. The evaluation of vulnerability is conducted with security evaluation scan results based on OWASP ZAP scanner. Finally, the security of the framework was examined via qualitative analysis.

4.5.1 Accessibility

Accessibility data was used to prove that the proposed authentication framework has the ability to satisfy Web Content Accessibility Guidelines (WCAG) 2.1 guidelines. These guidelines are in place to ensure a web-based application is accessible for everyone.

The Google Lighthouse accessibility tests executed against the virtual agent in the simulation. This uses a systematic check which includes a number of audits to ensure that service accessibility criteria meet the WCAG 2.1 guidelines. The checks are conducted for several configurations and metadata values set by the virtual agent simulation application.

Figure 7 shows the accessibility report generated with a score of 100, which represent that the authenticated virtual agent is accessible by a wider range of customers. However, Tollefsen and Ausland [83] argue that automated accessibility tools can comfortably detect about 50% of the accessibility criteria. Therefore, manual checks were conducted by the researcher for the latest version of the browsers Firefox, Chrome, Safari and Microsoft Edge check for the accessibility and no accessibility

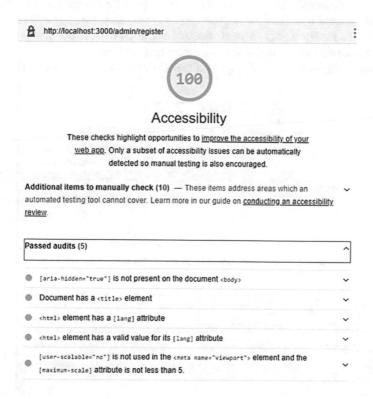

Fig. 7 Lighthouse accessibility reports

issues detected during these checks. Hence, the proposed authentication framework for virtual agents has the ability to comply with the WCAG guidelines.

4.5.2 Scalability

Scalability data is extracted from the simulation using Apache JMeter tests to prove that the authentication framework carries the ability to scale in practice without significant degradation of user performance. The concurrent authentication and de-authentication tests conducted for user samples of 100, 250 and 500 to demonstrate the scalability of the framework.

The infrastructure scalability tests are considered as out of scope for this demonstration due to time and cost constraints. These tests are conducted to evaluate the efficiency of scalability of the authentication process without consideration of MFA token generation or input time. Therefore, a static token is used to run the tests against the system. Evaluating the actual MFA token generation process and input time requires further work and it is not part of this study.

Figure 8 demonstrates the data of login and logout time in milliseconds (ms) for a 100 concurrent user sample. The test results provided an average of 21 ms for the login and 12 ms for logout. A total standard deviation of 7.24 ms demonstrates the number of deviations from the average authentication time.

Figure 9 shows the data of login and logout time in milliseconds (ms) for a 250

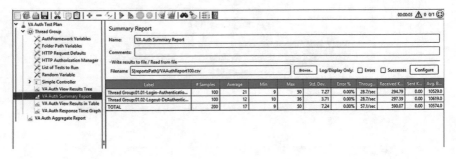

Fig. 8 Authentication/De-authentication performance test summary for 100 user sample

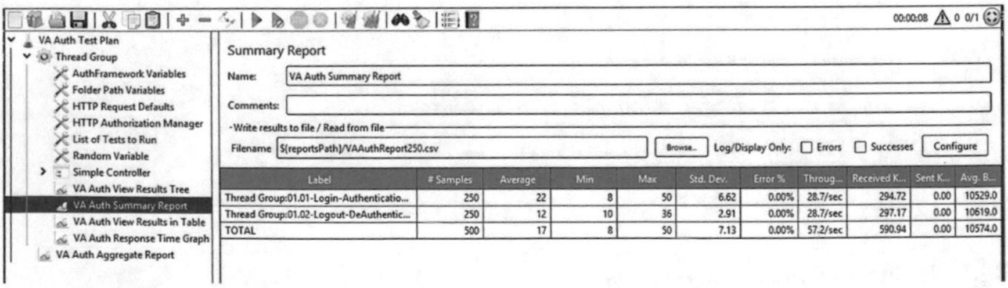

Fig. 9 Authentication/De-authentication performance test summary for 250 user sample

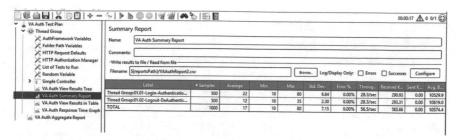

Fig. 10 Authentication/De-authentication performance test summary for 500 user sample

concurrent user sample. The execution of 250 user samples demonstrated an average of 22 ms for login performance and an average of 12 ms for logout. The total standard deviation has further reduced at this point and it is 7.13 ms.

Figure 10 demonstrates the test execution for 500 user samples with an average of 22 ms for login and 12 ms for logout. This is still in align with the 250 user samples and no difference is noticed in response time averages. The standard deviation total for 500 user login and logout attempts are at 7.15 ms.

The login and logout times obtained from the performance tests conducted for all three user samples demonstrated a consistent average for authentication time. This verifies the theoretical ability of the proposed framework to scale up without significant performance impact. The performance stats obtained are within the standard acceptable page load time below 2 s. Therefore, the authentication step does not have any adverse effects on user experience.

Standard deviation represents the number of test scenarios deviating from the average response time. The standard deviation for all three user sample scenarios is between 7.13 and 7.15 ms which represents stable performance over time. The maximum authentication time of 80 ms taken by any request over the three test scenarios illustrates fast and consistent authentication scalability with little variation.

The error rate of 0.00% represents the accuracy of the authentication and token handling process. This demonstrates that the authentication framework is capable of performing accurately with a 100% success rate within the given range of sample user scenarios.

4.5.3   Security

The key security properties are analysed qualitatively to ensure the security of the proposed authentication framework.

Table 2 illustrates the authentication security properties and respective security controls suggested with the framework.

The identity protection can be secured by using strong authentication mechanisms, MFA tokens and biometrics. As highlighted in the literature review, the proposed

Table 2 Key security properties and controls considered in the proposed authentication framework

Authentication security property	Security control
Identity protection of users	Strong authentication with MFA tokens from user devices and Biometrics
Login and logout	Tokens are invalidated immediately after the action
Data transfer	TLS encryption for all data transfers
API endpoints	Encrypted communication using certificates
Use of ACLs	Restrict the access of APIs to required sources
Monitoring of authentication process	Appropriate monitoring tools are configured to monitor the application and the infrastructure
WAF	Appropriate WAF configuration rules for top OWASP threats

framework provides the ability to leverage user devices to perform MFA or biometrics authentication to generate authentication tokens. A secured login and logout process require to ensure that no tokens can be re-used. Therefore, tokens have to be invalidated after being used immediately.

The encryption at transfer is performed with TLS 1.2 or higher to meet the current best encryption standard. Restricted API endpoints using ACLs provides limited access to the selected endpoints. Monitoring of authentication process provides insights about what is happening in the application and infrastructure. The adaption of an enterprise WAF solution can provide the mitigations against top OWASP threats.

Cyber-attack mitigation controls are introduced along with the authentication framework to ensure the ability to defend or mitigate cyber-attacks. As this research only focuses on the authentication, controls are considered as an overview based on mitigation of authentication attacks. A number of security controls such as MFA, rate-limiting of requests, honeypots, implementation of intrusion detection and prevention systems and WAF have been suggested to mitigate attacks against the authentication.

A research conducted to evaluate the authentication methods in the banks highlighted that adapting a strong authentication framework reduces the risk of cyber-attacks [5]. This highlights that the proposed authentication framework itself is providing the defences to mitigate the cyber-attacks against the authentication of virtual agents in the banks.

4.5.4 Vulnerability

The vulnerability data was obtained using the execution of OWASP ZAP Zed Attack proxy against API endpoint of the virtual agent. The purpose of this activity is to detect any vulnerabilities from the simulation environment as it is considered as an MVP environment for this research.

◊ ZAP Scanning Report

Summary of Alerts   **Generated**

Risk Level	Number of Alerts
High	0
Medium	1
Low	2
Informational	0

Alerts

Name	Risk Level	Number of Instances
X-Frame-Options Header Not Set	Medium	1
Incomplete or No Cache-control and Pragma HTTP Header Set	Low	2
X-Content-Type-Options Header Missing	Low	4

Fig. 11 OWASP ZAP vulnerability summary report

This activity analyses all the vulnerabilities related to the authentication framework from the external perspective. However, the previous researchers only conducted this test for a particular authentication method in a qualitative manner.

Figure 11 demonstrate the risk level and a number of alerts detected from the simulation environment. The vulnerability test conducted did not detect any high vulnerabilities in the environment. The X-Frame-Options alert detected was not applicable for this scenario as the authentication framework is only exposed via API calls. The auth API's contains JSON responses hence they do not render any embedded objects such as <iframe> or <frame>. The alert related to the cache-control was related to the fact that no-caching server is implemented with the simulation.

In a realistic scenario, it is possible to use elasticate or Redis utilities to cache certain authentication API calls. Therefore, three vulnerabilities detected are negligible and not applicable to this scenario. This demonstrates the ability of the framework to satisfy vulnerability characteristics.

4.6 Comparison of Results with Other Work

The results from the validation tests demonstrate that the proposed authentication framework complies with the evaluation criteria. This emphasis the suitability of the proposed authentication framework. The researcher has conducted a comparison of the results with other researchers who have proposed authentication frameworks in different industries to further clarify the findings.

The researcher conducted an authentication characteristic and performance comparison of existing authentication frameworks against the proposed framework.

Table 3 Authentication framework characteristics comparison

	C1	C2	C3	C4	C5	C6	C7
[18]	Yes	Yes	No	Yes	No	No	No
Kiljan et al. [36]	Yes	No	No	Yes	No	No	No
Liu et al. [46]	Yes	No	Yes	No	No	No	No
Proposed framework	Yes	Yes	Yes	Yes	Yes	Yes	Yes

C1: Type of credentials supports strong customer authentication
C2: Ability to manage credentials securely
C3: Monitoring of the authentication process
C4: High quality credentials support with the ability to mitigate tampering
C5: Authentication service supports OAuth2 integrations
C6: Information governance requirements are considered
C7: Compatibility for virtual agents considered

This comparison was conducted using the results from different researchers related to the authentication domain to understand the significance of the results of this research.

Table 3 demonstrates a comparison of the authentication characteristics of the proposed framework against prior related frameworks. The comparison reveals that all the studies have considered strong authentication in their respective frameworks. However, none of them has considered the adaptability of the frameworks to the virtual agents, OAuth2 integration support or governance requirements. Although, the proposed holistic authentication framework considers the above factors as they are necessary with the strict nature of the UK financial industry.

The researchers have not highlighted the process of managing the credentials securely in their authentication models except [18]. This may be due to the management of the credentials was not part of the scope of the studies. However, it is important to ensure that a solid process exists for the identity management of authentication framework suggested for the UK banks. Therefore, this study has ensured the management of the credentials is part of the scope of the proposed framework.

In summary, the authentication frameworks in comparison do not employ all the characteristics required for the authentication of virtual agents. However, the proposed framework meets all the required criteria.

The researcher also compares the authentication performance of different authentication frameworks in other industries to evaluate the findings of this research. This comparison was conducted qualitatively, as a quantitative comparison of results from different industries can be subjective.

Table 4 illustrates the authentication performance comparison of the proposed framework against other related frameworks. The comparison against five other researchers highlighted that only [36] has considered the accessibility criteria. Moreover, this research studied the accessibility from the customer perspective using a survey. Therefore, a systematic study of accessibility criteria was not part of their research. Hence, the researcher has conducted a systematic validation for the accessibility variable using the Google lighthouse utility.

Table 4 Authentication performance comparison

	P1	P2	P3	P4	P5
[18]	No	Yes	Yes	Yes	No
Nagaraju and Latha (2015)	No	Yes	Yes	Yes	No
Kiljan et al. [36]	Yes	No	Yes	Yes	No
Liu et al. [46]	No	Yes	Yes	No	No
Proposed framework	Yes	Yes	Yes	Yes	Yes

P1: Framework considers the accessibility
P2: Framework considers the scalability
P3: Framework considers the security
P4: Framework considers the vulnerability
P5: Framework considers the compliance

Furthermore, all the researchers in comparison did not consider the compliance requirements in their respective studies. As highlighted in this research, the UK e-banking applications require to comply with compliance standards such as open banking, PSD2 and FCA.

It was highlighted that only [46] has considered the monitoring of authentication. However, this research considers the monitoring of authentication process with the proposed holistic framework. Further, this study has highlighted the important monitoring elements required to maintain the auditability and compliance standards.

The comparison of scalability considerations against other related frameworks demonstrates greater variations in response time over the sample size. This was due to the fact that these studies did not focus on scalability from the stability or systematic perspective. Therefore, this study has occupied three sample sizes of 100, 250 and 500 to demonstrate that no significant performance degradation with the increased sample sizes. However, this is also linked with the infrastructure capacity and utilisation of the servers.

All the frameworks considered for the comparison consists of certain security considerations in their respective areas. However, none of them comprises of security considerations related to the financial services or virtual agents in the UK. This research has especially focused on the framework security consideration required by UK banking institutions.

Majority of the studies have considered attacks and vulnerabilities against the authentication methods except for Liu et al. [46]. In this research, the vulnerability is considered from the overall framework perspective instead of the particular authentication method. This provides the ability to dynamically adapt better authentication methods to the proposed framework. The vulnerability tests conducted in this research with OWASP ZAP pointing to the API endpoint covers overall external vulnerabilities of the application and the potential infrastructure configuration exposed.

In summary, the analysis of the virtual agents in major UK banks demonstrated that no current authentication framework exists for virtual agents. After an in-depth analysis, this study proposed a framework which covers all the requirements considered. The functionality of the proposed framework elaborated on how the framework functions within the banking environment.

Evaluation of the proposed framework consists of six core elements used as the guidelines and compliance of the framework. The analysis of the results of the experiments emphasises that the proposed framework meets the validation criteria. Finally, comparison of results with other researchers highlighted that the proposed framework has superior prospects against the authentication of the virtual agents.

Thus, this research fills the knowledge gap in the existing knowledge by introducing a suitable framework for the virtual agents in the UK banks which answers to the second research question of this study.

5 Conclusion and Future Works

5.1 Conclusions

This research was conducted to examine the authentication of virtual agents in the UK banking industry. The critical analysis of existing literature highlighted a lack of authentication of virtual agents in the UK banking industry. The majority of prior researchers conducted to evaluate the authentication of e-banking applications were focussed on user perspectives rather than a technical analysis. Furthermore, it was highlighted that existing e-banking authentication frameworks cannot be adapted to the virtual agents directly. Therefore, this research aimed at analysing the importance of the authentication of virtual agents and proposing a suitable authentication framework.

The existing literature on concepts such as virtual agents, AI and e-banking provided a solid foundation and better understanding to the first research question. Furthermore, the review has demonstrated the existing authentication models, frameworks and mechanisms in different industries are at risk of exposing sensitive customer data due to the lack of authentication. The knowledge gathered from current authentication methods and frameworks used in the other e-banking applications and open banking requirements helped to set the project in the right direction to progress with the methodology.

This research was based on sequential mixed methodology, where the qualitative analysis was followed by a systematic quantitative study. The data from previous research were used to conduct qualitative analysis to develop the proposed authentication framework. Similarly, a quantitative analysis was conducted to validate the proposed framework.

The validation of the framework was conducted using a simulation due to the restrictive nature of accessing the actual banking data. A simulation was configured and implemented in the researchers' AWS account to execute the validity tests. Google Lighthouse accessibility tests, Apache JMeter performance tests for three use cases and OWASP ZAP vulnerability tests were conducted against the simulated environment to extract the data required to demonstrate the accessibility, scalability and vulnerability of the proposed authentication framework.

The critical discussion carried out to investigate the authentication methods in the major UK banks demonstrated that the majority of them does not contain any authentication mechanism for their virtual agents. This verified the risk of compromisation of sensitive customer data by hackers and cyber-attacks. Hence, this study has proposed a holistic authentication framework for virtual agents in the UK banking industry. This proposed holistic authentication framework and functionalities elaborates the components and the behaviour of the authentication framework.

Adapting an authentication framework for UK banks required consideration of a number of different laws and standards such as open banking, PSD2, FCA, CBEST and data protection act. The core focus on the authentication framework from these standards is to ensure the CIA of the e-banking applications. The proposed authentication framework was designed based on this criterion. The design also demonstrated the low-level authentication process using a sequence diagram. Also, the proposed framework was designed to satisfy the six core elements suggested by GPG 44 and 45.

The analysis of the results of the validity variables (accessibility, scalability, vulnerability and security) shows that the proposed authentication framework is fit for the purpose. Furthermore, a qualitative security analysis was conducted from a high-level perspective to provide guidelines to secure the framework to mitigate the authentication attacks. Finally, the comparison conducted to compare the proposed authentication characteristics and authentication performance with other research highlighted the significant contribution of the proposed authentication framework to the existing knowledge.

5.2 Recommendations

The authentication helps to ensure the confidentiality and integrity of banking applications. Thus, the researcher recommends adapting a suitable authentication framework for virtual agents in UK banking institutions. This provides the ability to increase the trust of the existing customers as well as attract more customers to use virtual agents. Further, this offers the banks to utilise virtual agents profitably with enhanced e-banking functionalities.

The proposed framework provides high-level guidance on suitable password policy, monitoring policy, and infrastructure security policy to ensure the security of the entire authentication infrastructure. However, it is recommended to consider these elements in-depth in a practical implementation.

The proposed framework also has the ability to deploy as a standalone subsystem. Therefore, it is possible to consider the adaptability of the proposed authentication framework to other industries. The UK banking industry requires a higher grade of security. However, other industries may not require this level of security and complexity. The proposed authentication framework can be adapted with minor changes in such scenarios.

5.3   Limitations and Future Works

A production banking system can receive millions of authentication requests per second. Testing of such a large number of requests requires more computing power and time. As this research was conducted over 13 weeks period, the focus of this study was to prove that the propose authentication framework has the ability to scale while ensuring the accessibility, scalability, vulnerability and security standards. Therefore, careful consideration is required when generalising the results into a larger sample size.

Due to the strict security restrictions and limitations of access to the data, a simulation package is used to validate the authentication. However, this does not limit the ability to execute the tests in an actual authentication framework use in a bank or a financial organisation.

The practical authentication time consists of token generation time and input time which is a manual activity. Since this research uses a simulation environment, validating the MFA token generation and manual input time is not considered. The framework validation tests conducted uses a static token as MFA forbids the use of automation tests,

There are extremely limited studies conducted in the area of authentication of virtual agents. Therefore, more research can be conducted to evaluate the authentication of virtual agents in the financial industry as well as in other service industries.

The ability to collaborate e-banking application with popular virtual assistants such as Amazon Alexa, Google Assistant, Apple Siri, Microsoft Cortana will be an interesting area for further study. This can provide wider coverage for the virtual agents by making them accessible through the virtual agent preferred by the end-users.

References

1. Aliane L, Adda M (2019) Hobac: toward a higher-order attribute-based access control model. In: Procedia computer science. Elsevier B.V., pp 303–310. https://doi.org/10.1016/j.procs.2019. 08.044
2. Alsop T (2020) UK IT community views about changes to technology trends in future 2019. Statista, Statista. https://www.statista.com/statistics/1117521/it-community-views-about-cha nges-to-technology-trends/. Accessed 19 Nov 2020
3. Atlam HF, Wills GB (2019) An efficient security risk estimation technique for risk-based access control model for IoT. Internet of Things 6:100052. https://doi.org/10.1016/j.iot.2019.100052
4. Balla PB, Jadhao KT (2018) IoT based facial recognition security system. In: 2018 international conference on smart city and emerging technology, ICSCET 2018. Institute of Electrical and Electronics Engineers Inc. https://doi.org/10.1109/ICSCET.2018.8537344
5. Bani-Hani A, Majdalweieh M, AlShamsi A (2019) Online authentication methods used in banks and attacks against these methods. In: Procedia computer science. Elsevier B.V., pp 1052–1059. https://doi.org/10.1016/j.procs.2019.04.149
6. Bank of England (2016) CBEST intelligence-led testing understanding cyber threat intelligence operations. http://creativecommons.org/licenses/by/4.0/orsendalettertoCreativeC ommons. Accessed 3 Jan 2021
7. Barzu AP, Carabas M, Tapus N (2017) Scalability of a web server: how does vertical scalability improve the performance of a server? In: Proceedings—2017 21st international conference on control systems and computer, CSCS 2017. Institute of Electrical and Electronics Engineers Inc., pp 115–122. https://doi.org/10.1109/CSCS.2017.22
8. Bian W et al (2020) Bio-AKA: an efficient fingerprint based two factor user authentication and key agreement scheme. Futur Gener Comput Syst 109:45–55. https://doi.org/10.1016/j.future. 2020.03.034
9. Bonneau J et al (2012) The quest to replace passwords: a framework for comparative evaluation of web authentication schemes. In: Proccedings—IEEE symposium on security and privacy. Institute of Electrical and Electronics Engineers Inc., pp 553–567. https://doi.org/10.1109/SP. 2012.44
10. Brill TM, Munoz L, Miller RJ (2019) Siri, Alexa, and other digital assistants: a study of customer satisfaction with artificial intelligence applications. J Mark Manag 35(15–16):1401–1436. https://doi.org/10.1080/0267257X.2019.1687571
11. Cabrera-Sánchez J-P et al (2020) Identifying relevant segments of AI applications adopters—expanding the UTAUT2's variables. Telemat Informat, p 101529. https://doi.org/10.1016/j.tele. 2020.101529
12. Cai H, Venkatasubramanian KK (2017) Patient identity verification based on physiological signal fusion. In: Proceedings—2017 IEEE 2nd international conference on connected health: applications, systems and engineering technologies, CHASE 2017. Institute of Electrical and Electronics Engineers Inc., pp 90–95. https://doi.org/10.1109/CHASE.2017.65
13. Caldwell T (2017) HSBC reports high trust levels in biometric tech as twins spoof its voice ID system. Elsevier Enhanced Reader. https://reader.elsevier.com/reader/sd/pii/S09694765173 01194?token=AAB9E90853B7CB7A309603052E2E8410D404A9B0DF10279CCF592A 0C19991ADE375881B12EED275AB4CF21F2E753E90F. Accessed 23 Nov 2020
14. Chubarov AA et al (2020) Virtual listener: a turing-like test for behavioral believability. In: Procedia computer science. Elsevier B.V., pp 892–899. https://doi.org/10.1016/j.procs.2020. 02.146
15. Conrad E, Misenar S, Feldman J (2014) Domain 1: access control. In: Eleventh hour CISSP. Elsevier, pp 1–21. https://doi.org/10.1016/b978-0-12-417142-8.00001-7
16. Crosman P (2018) How AI virtual assistants (like Alexa) could reshape mortgages: an AI-powered virtual assistant could be used in a variety of ways, including helping customers to prequalify for mortgages, easing compliance and detecting problems. Nat Mortgage News 43(3):N.PAG-N.PAG. http://search.ebscohost.com/login.aspx?direct=true&db=buh&AN=132 389570&authtype=shib&site=ehost-live&scope=site

17. Dan A et al (2019) Toward an AI chatbot-driven advanced digital locker. In: Advances in intelligent systems and computing. Springer, pp 37–46. https://doi.org/10.1007/978-981-13-1544-2_4
18. Darwish SM, Hassan AM (2012) A model to authenticate requests for online banking transactions. Alex Eng J 51(3):185–191. https://doi.org/10.1016/j.aej.2012.02.005
19. Devi RS, Kumar MM (2020) Testing for security weakness of web applications using ethical hacking. In: Proceedings of the 4th international conference on trends in electronics and informatics, ICOEI 2020. Institute of Electrical and Electronics Engineers Inc., pp 354–361. https://doi.org/10.1109/ICOEI48184.2020.9143018
20. Dobbie S (2020) Challenge of biometric security for banks. Biometric Technol Today 3:5–7. https://doi.org/10.1016/S0969-4765(20)30037-0
21. Dodanduwa K, Kaluthanthri I (2018) Role of trust in OAuth 2.0 and OpenID connect. In: 2018 IEEE 9th international conference on information and automation for sustainability, ICIAfS 2018. Institute of Electrical and Electronics Engineers Inc. https://doi.org/10.1109/ICIAFS.2018.8913384
22. Ethelbert O et al (2017) A JSON token-based authentication and access management schema for cloud SaaS applications. In: Proceedings—2017 IEEE 5th international conference on future internet of things and cloud, FiCloud 2017. Institute of Electrical and Electronics Engineers Inc., pp 47–53. https://doi.org/10.1109/FiCloud.2017.29
23. FCA (2020) Strong customer authentication. FCA, FCA UK. https://www.fca.org.uk/firms/str ong-customer-authentication. Accessed 5 Jan 2021
24. FCA UK (2020) Strong customer authentication. FCA, FCA. https://www.fca.org.uk/consum ers/strong-customer-authentication. Accessed 12 Jan 2021
25. Friesen M et al (2020) A comparative evaluation of security mechanisms in DDS, TLS and DTLS, pp 201–216. https://doi.org/10.1007/978-3-662-59895-5_15
26. Følstad A, Nordheim CB, Bjørkli CA (2018) What makes users trust a chatbot for customer service? An exploratory interview study. In: Lecture notes in computer science (including subseries lecture notes in artificial intelligence and lecture notes in bioinformatics). Springer, pp 194–208. https://doi.org/10.1007/978-3-030-01437-7_16
27. Garín-Muñoz T et al (2019) Models for individual adoption of eCommerce, eBanking and eGovernment in Spain. Telecommun Policy 43(1):100–111. https://doi.org/10.1016/j.telpol.2018.01.002
28. Guo C et al (2018) Fraud risk monitoring system for e-banking transactions. In: 2018 IEEE 16th international conference on dependable, autonomic and secure computing, 16th international conference on pervasive intelligence and computing, 4th international conference big data intelligence and computing and cyber science and technology congress (DASC/PiCom/DataCom/CyberSciTech), pp 100–105. https://doi.org/10.1109/DASC/PiCom/DataCom/CyberSciTec.2018.00030
29. Haekal M, Eliyani (2017) Token-based authentication using JSON Web Token on SIKASIR RESTful Web Service. In: 2016 international conference on informatics and computing, ICIC 2016. Institute of Electrical and Electronics Engineers Inc., pp 175–179. https://doi.org/10.1109/IAC.2016.7905711
30. Hassan T, Ahmed F (2019) Transaction and identity authentication security model for e-banking: confluence of quantum cryptography and AI. SpringerLink transaction and identity authentication security model for e-banking: confluence of quantum cryptography and AI international conference on intelligent technologies and applications INTAP 2018: Intelligent Technologies and Applications, pp 338–347. Cite as', Communications in computer and information science. https://doi.org/10.1007/978-981-13-6052-7_29.
31. He Y, Sun H, Feng H (2020) UA-Miner: deep learning systems for expose unprotected API vulnerability in source code. In: 12th international conference on advanced computational intelligence, ICACI 2020. Institute of Electrical and Electronics Engineers Inc., pp 378–384. https://doi.org/10.1109/ICACI49185.2020.9177528
32. Huang X, Zhang Y (2020) Indistinguishability and unextractablility of password-based authentication in blockchain. Futur Gener Comput Syst 112:561–566. https://doi.org/10.1016/j.future.2020.05.009

33. Indu I, Rubesh Anand PM, Bhaskar V (2017) Encrypted token based authentication with adapted SAML technology for cloud web services. J Netw Comput Appl. Academic Press, pp 131–145. https://doi.org/10.1016/j.jnca.2017.10.001

34. Juniper Research (2018) Press releases from juniper research, Google Scholar. https://www.juniperresearch.com/press/press-releases. Accessed 30 Nov 2020

35. Kaiqb et al (2019) Virtual assistant overview—Bot Service. Microsoft Docs, Microsoft Docs. https://docs.microsoft.com/en-us/azure/bot-service/bot-builder-virtual-assistant-introduction?view=azure-bot-service-4.0. Accessed 18 Nov 2020

36. Kiljan S, Vranken H, van Eekelen M (2018) Evaluation of transaction authentication methods for online banking. Futur Gener Comput Syst 80:430–447. https://doi.org/10.1016/j.future.2016.05.024

37. Kolluru KK et al (2018) An AAA solution for securing industrial IoT devices using next generation access control. In: Proceedings—2018 IEEE industrial cyber-physical systems, ICPS 2018. Institute of Electrical and Electronics Engineers Inc., pp 737–742. https://doi.org/10.1109/ICPHYS.2018.8390799

38. Kritikos K et al (2019) A survey on vulnerability assessment tools and databases for cloud-based web applications. Array 3–4:100011. https://doi.org/10.1016/j.array.2019.100011

39. Krämer NC et al (2018) Social snacking with a virtual agent—on the interrelation of need to belong and effects of social responsiveness when interacting with artificial entities. Int J Hum Comput Stud 109:112–121. https://doi.org/10.1016/j.ijhcs.2017.09.001

40. Kumar V, Ramachandran D, Kumar B (2020) Influence of new-age technologies on marketing: a research agenda. J Bus Res. https://doi.org/10.1016/j.jbusres.2020.01.007

41. Kumar RG, Priya (2015) Multi-touch authentication framework for cloud secrecy over hidden data. In: 2015 2nd international conference on computing for sustainable global development (INDIACom), pp 64–66

42. Königstorfer F, Thalmann S (2020) Applications of artificial intelligence in commercial banks—a research agenda for behavioral finance. J Behav Exp Financ 27:100352. https://doi.org/10.1016/j.jbef.2020.100352

43. Lenka RK et al (2018) Performance analysis of automated testing tools: JMeter and TestComplete. In: Proceedings—IEEE 2018 international conference on advances in computing, communication control and networking, ICACCCN 2018. Institute of Electrical and Electronics Engineers Inc., pp 399–407. https://doi.org/10.1109/ICACCCN.2018.8748521

44. Leung L (2015) Validity, reliability, and generalizability in qualitative research. J Family Medi Primary Care 4(3):324. https://doi.org/10.4103/2249-4863.161306

45. Leung J, Shen Z, Miao C (2019) Goal-oriented modelling for virtual assistants. In: Proceedings—2019 IEEE international conference on agents, ICA 2019. Institute of Electrical and Electronics Engineers Inc., pp 73–76. https://doi.org/10.1109/AGENTS.2019.8929177

46. Liu W, Uluagac AS, Beyah R (2014) MACA: a privacy-preserving multi-factor cloud authentication system utilizing big data. In: Proceedings—IEEE INFOCOM. Institute of Electrical and Electronics Engineers Inc., pp 518–523. https://doi.org/10.1109/INFCOMW.2014.6849285

47. Lupu C, Gaitan VG, Lupu V (2015) Security enhancement of internet banking applications by using multimodal biometrics. In: SAMI 2015—IEEE 13th international symposium on applied machine intelligence and informatics, proceedings. Institute of Electrical and Electronics Engineers Inc., pp 47–52. https://doi.org/10.1109/SAMI.2015.7061904

48. Makino Y, Klyuev V (2015) Evaluation of web vulnerability scanners. In: Proceedings of the 2015 IEEE 8th international conference on intelligent data acquisition and advanced computing systems: technology and applications, IDAACS 2015. Institute of Electrical and Electronics Engineers Inc., pp 399–402. https://doi.org/10.1109/IDAACS.2015.7340766

49. McDowell B (2019) Three ways in which GDPR impacts authentication. Comput Fraud Secur 2:9–12. https://doi.org/10.1016/S1361-3723(19)30019-3

50. McLean G, Osei-Frimpong K, Barhorst J (2021) Alexa, do voice assistants influence consumer brand engagement?—examining the role of AI powered voice assistants in influencing consumer brand engagement. J Bus Res 124:312–328. https://doi.org/10.1016/j.jbusres.2020.11.045

51. Mihajlov M, Blažič BJ, Josimovski S (2011) Quantifying usability and security in authentication. In: Proceedings—international computer software and applications conference, pp 626–629. https://doi.org/10.1109/COMPSAC.2011.87
52. Moar J (2019) Hey siri, how will you make money? Whitepapers, Juniper Research. https://www.juniperresearch.com/document-library/white-papers/hey-siri-how-will-you-make-money. Accessed 17 Nov 2020
53. Mogaji E, Danbury A (2017) Making the brand appealing: advertising strategies and consumers' attitude towards UK retail bank brands. J Prod Brand Manage 26(6):531–544. https://doi.org/10.1108/JPBM-07-2016-1285
54. Morgen B (2017) 5 ways Chatbots can improve customer experience in banking, Forbes. https://www.forbes.com/sites/blakemorgan/2017/08/06/5-ways-chatbots-can-improve-customer-experience-in-banking/?sh=1867066f7148. Accessed 16 Jan 2021
55. Nadarzynski T et al (2019) Acceptability of artificial intelligence (AI)-led chatbot services in healthcare: a mixed-methods study. Digital Health 5:205520761987180. https://doi.org/10.1177/2055207619871808
56. Naik N (2016) Connecting google cloud system with organizational systems for effortless data analysis by anyone, anytime, anywhere. In: ISSE 2016—2016 international symposium on systems engineering—proceedings papers. Institute of Electrical and Electronics Engineers Inc. https://doi.org/10.1109/SysEng.2016.7753150
57. Nami MR (2009) E-banking: issues and challenges. In: 10th ACIS conference on software engineering, artificial intelligence, networking and parallel/distributed computing, SNPD 2009, In conjunction with IWEA 2009 and WEACR 2009, pp 263–266. https://doi.org/10.1109/SNPD.2009.60
58. Navas J, Beltrán M (2019) Understanding and mitigating OpenID Connect threats. Comput Secur 84:1–16. https://doi.org/10.1016/j.cose.2019.03.003
59. Nuseibah A et al (2017) Minimum viable product creation through adaptive project management—an extended approach for the management of innovation projects: the ecochallenge case. In: Proceedings of the 2017 IEEE 9th international conference on intelligent data acquisition and advanced computing systems: technology and applications, IDAACS 2017. Institute of Electrical and Electronics Engineers Inc., pp 446–452. https://doi.org/10.1109/IDAACS.2017.8095121
60. Perlines Hormann T, Wrona K, Holtmanns S (2006) Evaluation of certificate validation mechanisms. Comput Commun 29(3):291–305. https://doi.org/10.1016/j.comcom.2004.12.008
61. Pradeep S, Sharma YK (2019) A pragmatic evaluation of stress and performance testing technologies for web based applications. In: Proceedings—2019 amity international conference on artificial intelligence, AICAI 2019. Institute of Electrical and Electronics Engineers Inc., pp 399–403. https://doi.org/10.1109/AICAI.2019.8701327
62. Prapty RT et al (2020) Preventing session hijacking using encrypted one-time-cookies. In: Wireless telecommunications symposium. IEEE Computer Society. https://doi.org/10.1109/WTS48268.2020.9198717
63. Premchand A, Choudhry A (2018) Open banking & APIs for transformation in banking. In: 2018 international conference on communication, computing and internet of things (IC3IoT), pp 25–29. https://doi.org/10.1109/IC3IoT.2018.8668107
64. Premchand A, Choudhry A (2019) Open banking and APIs for transformation in banking. In: Proceedings of the 2018 international conference on communication, computing and internet of things, IC3IoT 2018. Institute of Electrical and Electronics Engineers Inc., pp 25–29. https://doi.org/10.1109/IC3IoT.2018.8668107
65. Puthal D et al (2019) Secure authentication and load balancing of distributed edge datacenters. J Para Distribut Comput 124:60–69. https://doi.org/10.1016/j.jpdc.2018.10.007
66. Pérez-Méndez A et al (2013) Out-of-band federated authentication for Kerberos based on PANA. Comput Commun 36(14):1527–1538. https://doi.org/10.1016/j.comcom.2013.07.004
67. Reddy T (2017) How chatbots can help reduce customer service costs by 30%—Watson Blog, IBM. https://www.ibm.com/blogs/watson/2017/10/how-chatbots-reduce-customer-service-costs-by-30-percent/. Accessed 18 Nov 2020

68. Renaud K (2004) Quantifying the quality of web authentication mechanisms: a usability perspective. J Web Eng—JWE
69. Rensing C et al (2002) AAA: a survey and a policy-based architecture and framework. IEEE Netw 16(6):22–27. https://doi.org/10.1109/MNET.2002.1081762
70. Rese A, Ganster L, Baier D (2020) Chatbots in retailers' customer communication: how to measure their acceptance? J Retail Consum Serv 56:102176. https://doi.org/10.1016/j.jretco nser.2020.102176
71. Richer J, Sanso A (2017) OAuth 2 in action, 1st edn. Manning Publications. https://www.man ning.com/books/oauth-2-in-action. Accessed 3 Jan 2021
72. Rivera Sánchez YK, Demurjian SA, Baihan MS (2019) A service-based RBAC & MAC approach incorporated into the FHIR standard. Digital Commun Netw 5(4):214–225. https:// doi.org/10.1016/j.dcan.2019.10.004
73. Sadikin F, van Deursen T, Kumar S (2020) A ZigBee intrusion detection system for IoT using secure and efficient data collection. Internet of Things 12:100306. https://doi.org/10.1016/j. iot.2020.100306
74. Saunders MNK, Lewis P, Thornhill A (2015) Research methods for business students. Pearson Education UK, Harlow, United Kingdom
75. Schydlowsky DM (2020) Prudential regulations for greening the financial system: coping with climate disasters. Latin Am J Central Bank 1(1–4):100010. https://doi.org/10.1016/j.latcb. 2020.100010
76. Sendor J et al (2014) Platform-level support for authorization in cloud services with OAuth 2. In: Proceedings—2014 IEEE international conference on cloud engineering, IC2E 2014. Institute of Electrical and Electronics Engineers Inc., pp 458–465. https://doi.org/10.1109/ IC2E.2014.60
77. Sheerman L et al (2020) COVID-19 and the secret virtual assistants: the social weapons for a state of emergency. Emerald Open Res 2:19. https://doi.org/10.35241/emeraldopenres.135 71.1
78. Soderlund M, Oikarinen EL, Tan TM (2020) The happy virtual agent and its impact on the human customer in the service encounter. J Retail Consum Serv 59:102401. https://doi.org/10. 1016/j.jretconser.2020.102401
79. Solapurkar P (2016) Building secure healthcare services using OAuth 2.0 and JSON web token in IOT cloud scenario. In: Proceedings of the 2016 2nd international conference on contemporary computing and informatics, IC3I 2016. Institute of Electrical and Electronics Engineers Inc., pp 99–104. https://doi.org/10.1109/IC3I.2016.7917942
80. Statista (2020) AI chatbots: usage in customer service organizations 2018. Statista, Statista. https://www.statista.com/statistics/1025267/ai-chatbots-usage-in-customer-service-organizations/. Accessed 19 Nov 2020
81. Statista Research Department (2020) Chatbot ability in customer service U.S., Canada & U.K. 2019. Statista. https://www.statista.com/statistics/1015841/customer-service-chatbot-abi lity-us-canada-uk/. Accessed 19 Nov 2020
82. Sunardi A, Suharjito (2019) MVC architecture: a comparative study between laravel framework and slim framework in freelancer project monitoring system web based. In: Procedia computer science. Elsevier B.V., pp 134–141. https://doi.org/10.1016/j.procs.2019.08.150
83. Tollefsen M, Ausland T (2018) A practitioner's approach to using WCAG evaluation tools. In: 2017 6th international conference on information and communication technology and acces-sibility, ICTA 2017. Institute of Electrical and Electronics Engineers Inc., pp 1–5. https://doi. org/10.1109/ICTA.2017.8336047
84. Tolosana R et al (2020) Deepfakes and beyond: a survey of face manipulation and fake detection. Inf Fus 64:131–148. https://doi.org/10.1016/j.inffus.2020.06.014
85. Trescak T, Bogdanovych A (2018) Simulating complex social behaviours of virtual agents through case-based planning. Comput Graph (pergamon) 77:122–139. https://doi.org/10.1016/ j.cag.2018.10.004
86. Tulshan AS, Dhage SN (2019) Survey on virtual assistant: Google assistant, Siri, Cortana, Alexa. In: Communications in computer and information science. Springer, pp 190–201. https:// doi.org/10.1007/978-981-13-5758-9_17

87. Tuza SKA (2019) Secure chatbots against data leak and over—learning threats. Google Scholar. http://repository.nauss.edu.sa//handle/123456789/66125. Accessed 19 Nov 2020
88. Urien P, Pasquet M, Kiennert C (2011) A breakthrough for prepaid payment: End to end token exchange and management using secure SSL channels created by EAP-TLS smart cards. In: Proceedings of the 2011 international conference on collaboration technologies and systems, CTS 2011, pp 476–483. https://doi.org/10.1109/CTS.2011.5928726
89. Vassilev V et al (2020) Two-factor authentication for voice assistance in digital banking using public cloud services. In: Proceedings of the confluence 2020—10th international conference on cloud computing, data science and engineering. Institute of Electrical and Electronics Engineers Inc., pp 404–409. https://doi.org/10.1109/Confluence47617.2020.9058332.
90. Voris J et al (2019) Active authentication using file system decoys and user behavior modeling: results of a large scale study. Comput Secur 87:101412. https://doi.org/10.1016/j.cose.2018. 07.021
91. Wagner TD et al (2019) Cyber threat intelligence sharing: survey and research directions. Comput Secur 87:101589. https://doi.org/10.1016/j.cose.2019.101589
92. Wille K, Dumke RR, Wille C (2017) Measuring the accessability based on web content accessibility guidelines. In: Proceedings—26th international workshop on software measurement, IWSM 2016 and the 11th international conference on software process and product measurement, Mensura 2016. Institute of Electrical and Electronics Engineers Inc., pp 164–169. https:// doi.org/10.1109/IWSM-Mensura.2016.032
93. Wojewidka J (2020) The deepfake threat to face biometrics. Biometric Technol Today 2:5–7. https://doi.org/10.1016/S0969-4765(20)30023-0
94. Xue M et al (2020) LOPA: a linear offset based poisoning attack method against adaptive fingerprint authentication system. Comput Secur 99:102046. https://doi.org/10.1016/j.cose. 2020.102046
95. Yang H (2019) EC cryptography tutorials—Herong's tutorial examples, 1st edn. Lulu.com. https://www.ebooks.com/en-uk/book/209712101/ec-cryptography-tutorials-herong-s-tut orial-examples/herong-yang/?_c=1. Accessed 4 Jan 2021
96. Younis YA, Kifayat K, Merabti M (2016) A novel evaluation criteria to cloud based access control models. In: Proceedings—2015 11th International Conference on Innovations in Information Technology, IIT 2015. Institute of Electrical and Electronics Engineers Inc., pp 68–73. https://doi.org/10.1109/INNOVATIONS.2015.7381517

Intelligent Security: Applying Artificial Intelligence to Detect Advanced Cyber Attacks

Karthik Kallepalli and Umair B. Chaudhry

Abstract Every organization and individual wants their system and network to be secure and not compromised and loose sensitive data. For this safety to be achieved, a firewall is incorporated in every computer/system. This firewall acts as a filter between the internet and the local computer/system which restricts unauthorized access and malicious content according to the rules embedded in the firewall's software. But, the hackers still managed to bypass these with few techniques using vulnerabilities of the firewall. Over a period of time, the firewalls are upgraded according to the attacks happening and the hackers approach them with new bypassing techniques. Therefore, while some can invest any amount on an updated and strong firewall, others cannot afford it, as the price vary according to the number of features the firewall offers. The latest upgrade of the firewall over the traditional firewall is Next-Generation firewall which does deep-packet inspection in the traffic. However, Next-Generation Firewall is also static in nature since it has to follow signatures. This chapter discusses possibly a new technique which may be more efficient than the Next-Generation Firewall and achieves far more protection than any other firewall on a '**WHAT IF**' basis. The new technique is allowing an 'Adaptive nature' to the firewall with the help of Artificial intelligence connected to the sandbox of the firewall. The adaptive nature studies the behavior of the traffic/anomalies while also following the static rules like traditional firewall. This has more security authentications and therefore provides higher security to both, an individual, as well as to an organization.

Keywords Artificial intelligence · Adaptive technology · Adaptive firewall · Intelligent firewall · Firewall with AI

K. Kallepalli · U. B. Chaudhry (✉)
Northumbria University London, London, UK
e-mail: Umair.chaudhry@northumbria.ac.uk

© The Author(s), under exclusive license to Springer Nature Switzerland AG 2021
R. Montasari et al. (eds.), *Challenges in the IoT and Smart Environments*,
Advanced Sciences and Technologies for Security Applications,
https://doi.org/10.1007/978-3-030-87166-6_11

1 Introduction

Security is a necessity to everyone from a personal computer in a home network to a conglomerate network. The entire network is protected by a firewall. The different kinds of firewalls are packet-filtering firewall, cloud firewall, software & hardware firewalls, Unified threat management (UTM) firewall, Stateful and stateless firewall and so on, where each one is used according to the requirement. These firewalls are regularly checked for faults and if any vulnerabilities are found, then the vendor of the respective firewall releases a patch where in most of the cases the firewall gets updated by the vendor via internet itself. Despite these many measures have been taken, the attacks are still taking place and the reason is, the firewalls are taking actions according to the rules given to them which in other words called 'Static Behavior'. Almost all the firewalls those are available in today's market are static, except for Next-Generation Firewall (NGF) provided by the vendors such as Palo Alto Networks, Barracuda Networks, Jupiter networks, Watchguards and so on. This Next-Generation Firewall (NGFW) has integrated 'Intrusion Prevention System' and 'Intrusion Detection System' which is not present in standard firewalls. NGFW does deep-packet filtering and detects malicious traffic in all the OSI layers. Though NGFWs have many unique features than the traditional firewalls; they also run on signature-based rules which make them somewhat static in nature. Now, we propose 'Adaptive nature' to a firewall which surpasses the NGFW in terms of prevention and detection based on the behavioral pattern. This is achieved through artificial intelligence that's connected to the sandbox of the firewall and the security is provided to the user/client through two-factor authentication as well as cloud-authentication when an anomaly occurs, which we will discuss as below.

2 Background

This chapter discusses the history and importance of the firewalls in personal systems and organizations, explaining its significance in the present day. A new technique called 'Artificial Intelligence' is proposed as a new development that is added to the current system of firewalls that enhances the performance in a more affective way.

3 What is a Firewall?

A firewall is a mechanism which implements two networks, such as the private LAN and the insecure, open Internet, with an access control policy. The firewall decides which services inside, and vice versa, can be accessed from the outside. The exact means with which this is done differ greatly, but the firewall should be considered as a pair of structures in principle: one that would hold up/block the traffic, and other one to accept/permit traffic. A firewall can not only consider as your network's locked front door but also as network's security guard. Firewalls are indeed significant because they have a centralized "choke point" where it is easy to enforce protections and audits. A firewall gives the information of the type and quantity of the traffic to the administrator of the system which has gone through that respective system, as well as the logs of the attempts which try to break through the system and so on. Your firewall not only blocks entry, but also tracks who are sniffing around, like a closed circuit surveillance TV device, and helps detect those who want to break your security. The primary purpose of a firewall basically is to protect the network by focusing on three things explained in [1].

- It blocks the incoming information that could involve an attack by a hacker.
- By making it appear like all outgoing traffic comes from the firewall rather than the network, it masks information about the network. This concept is known as Network Address Translation (NAT).
- In order to control Internet use and/or connections to remote locations, it screens outgoing traffic. Incoming and outgoing traffic may be screened by a firewall. It is typically screened more carefully than outgoing traffic because incoming traffic presents a greater danger to the network.

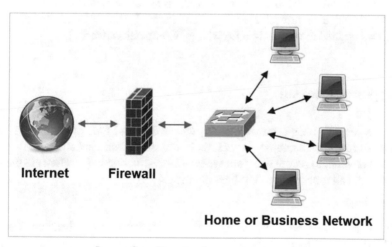

Internet   Firewall

Home or Business Network

Image Src: Practicalcomputersinc.com

Screening Levels:

You will track both incoming and outgoing traffic with a firewall. Because incoming traffic faces a higher network threat, it is typically inspected more specifically than outgoing traffic. You can typically learn the three ways of scanning that firewalls conduct when you look at firewall hardware or software devices.

- Screening which prevents/blocks all incoming information that a user does not explicitly order on the network
- Screening from the sender's address
- Screening of the content of the message.

Think about screening stages as an exclusion process. First, the firewall decides if the incoming transmission is anything that a person on the network requests, and ignores something more. Anything that is permitted in is then more carefully investigated. To guarantee that it is a trustworthy location, the firewall scans the sender's computer address. It reviews the contents of the transmission as well [1].

The essential feature of a firewall is to shield the system from an unauthorized user. The mechanism of filtering such packets is accomplished by a stable policy by filtering the packets passing it. A firewall plays a role in a computer network that is somewhat close to a firewall in a home. Just as one section of a house is covered by a shield built of concrete, a firewall in a network means that computers on the other side will not be affected if anything bad occurs on one side of the firewall. A network firewall must defend against several various forms of attacks, such as worms, viruses, DoS, unlike a building firewall, which defends against such a particular danger (fire) in [2].

A defense network which is intended to defend computer systems and networks from malware threats is a firewall. Antivirus, on the other hand, is a software utility program intended to protect a device from internal virus attacks. Another word sometimes applied in respect to a firewall is 'Computer network' [3].

4   Types of Firewall

There are several specific firewall types, where different type operates in a varied manner, in inside data centers as well as in organizational exclusion zones and in the cloud beyond, to secure different types of infrastructure in [4]. Here are the most relevant firewall types we need to hear about:

Packet Filtering Firewalls:

Packet filters are the type where the firewall searches for the packet headers to analyze them by comparing them to the ACLs aka Access Control Lists (ACLs), identified by network protection group [5]. In order to check whether the packet is safe to allow it to stay inside a network, the respective firewall extracts the information which is placed inside packet header, for instance port number and IP address. If the packet fails to set the set conditions for this firewall type, it will be dropped and will not be able to move through the network [5].

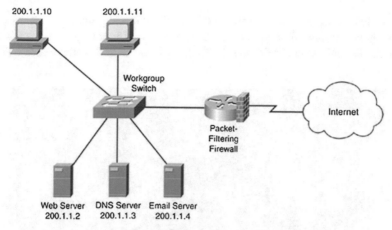

Image Src: nhprice.com

Circuit-Level Gateway Firewall:

These are the firewalls that are used for filtering the traffic that is flowing between a trustworthy inner host and an un-trusted external host. The key aim is to make sure that all the packets are in proper fashion which is involved in creating and controlling the connection/session between two hosts. If the link is established, no further packet monitoring will be needed. These kinds of session layer (Layer 5) or network layer (Layer 3) firewalls are used in the OSI model layer described in [6].

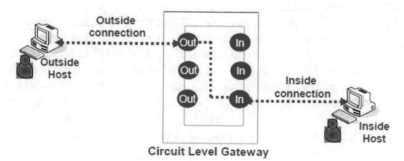

Image Src: bp.blogspot.com

Stateful inspection Firewall:

Whenever a firewall is defined as a Stateful inspection, it implies that this just checks network layer packets as packet filtering does, but instead of merely applying basic filtering standards to this content, it utilizes it to block unwanted traffic in an intelligent way. This analyzes details to ensure that link requests appear in the right format. It also controls each contact session from start to finish and enforces rules derived from protocol, source and destination addresses and port. Stateful inspection firewall will easily check that new incoming packets follow the requirements for permitted traffic by preserving all session data. Packets not used in an approved session are refused. The main feature of Stateful firewall is it can be both smart and fast [7].

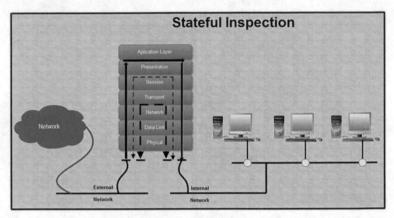

Image Src: ecomputernotes.com

Application-Level Gateways/Proxy server firewalls:

By implementing filtering and logging, authentication, Proxy enables clients to use a particular service (TELNET, FTP, HTTP, etc.) or connection type. There may be a separate proxy for specific services. For instance, if you want to give your internal network users only an HTTP access to the Internet, then you need to permit just a HTTP proxy but not anything more. Clients who want to access the Internet establish a proxy server along with a virtual circuit which initiates a request on behalf of the individual user to connect to a specific site. The proxy server switches the IP that belongs to that specific request such that only IP of that proxy server is accessible to the outside world or the Internet. The internal network is thus hidden behind the proxy server. Whenever a proxy collects online data, it transfers the data through the simulated circuit reverse to predestined internal customer. The primary benefit of using proxy is that it is completely aware of and can provide protection to the type of data it manages [8].

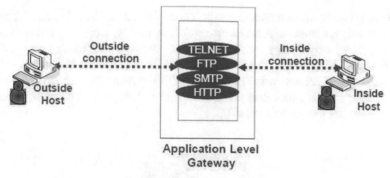

Application Level Gateway

Image Src: bp.blogspot.com

Hardware Firewalls:

In [9] *Eric Dosal, 2019*, says that Hardware firewalls choose a physical device that intercepts data packets and traffic requests in a manner equivalent to a traffic router until they are connected to the servers of the network. Hardware firewalls choose a device (physical) which requests the traffic and intercepts the data packets in a manner that is equivalent to the nature of a traffic router until they establish a connection with the network servers. Firewalls such as these physical-based appliances, succeed at the perimeter protection. They make sure that the unauthorized activity is identified outward the network until the company's network endpoints are informed about the vulnerabilities. However, a greatest drawback of this firewall is, it is sometimes convenient for insider attacks to penetrate the firewall. Often, depending on vendor, the exact specifications of a hardware firewall can vary; some may have a more restricted ability than others to accommodate simultaneous connections.

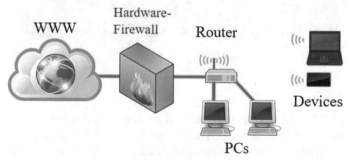

Image Src: pickmcablemodem.com

Software firewall:

The firewall is built on a typical computer for this form of firewall. The firewall runs on the computer as application software. Provided that this firewall accesses computer services via the base system in a shared network, it has less speed and consistency when compared to the hardware firewall. It has minimal functionality and functions, unlike a hardware firewall. Firewalls for software are easy to install. These firewalls can be conveniently configured by an ordinary network user to suit their security specifications. Since software firewalls need no extra hardware to operate, they do not lift the cost of the network. The benefits of the software firewall are low cost and simpler setup, while slow latency, less precision and the lack of extra functionality are the drawbacks explained in [10].

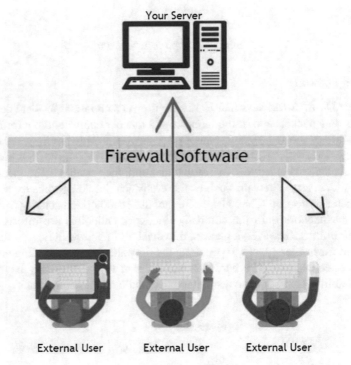

Image Src: Spiralmedia.com

Cloud Firewall:

With the additional bonus of low-cost maintenance, these firewalls are more like fusion of hardware and software. Though the hardware is built on the respective network, controlling the firewall is remotely done by vendor's cyber specialists you appoint. Many of the upgrades can be done automatically off-site, and then you can probably improve your infrastructure to meet a growing company's demands. A stable and advantageous Service Level Agreement (SLA) that stipulates 99.99% compatibility and responsiveness that complies with universal SOC guidelines is the downside of a cloud firewall [11].

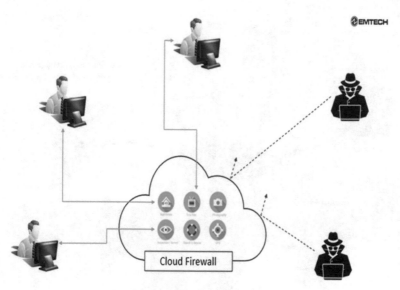

Image Src: emtech.ae

Unified threat management (UTM) firewall:

As per Cisco, 2020 in [12] a UTM firewall system usually integrates the roles of a Stateful inspection firewall with antivirus and intrusion protection, in a loosely coupled fashion. Additional resources and occasionally cloud management can also be used. Simplicity and ease of use are the priorities of UTMs. As a single box that links to the network, an almost full security solution is provided to companies like small, medium sized by these UTM devices. Standard firewalls, an intrusion prevention system (such as virus and malware e-mail, testing inward traffic, blacklisting), and a checklist/blacklist of domain names/Websites, are common UTM features to deter staff from visiting known websites, such as phishing. Stable Network gateways also feature the web device firewall and next-generation firewall (NGFW) functions (sometimes) in [13].

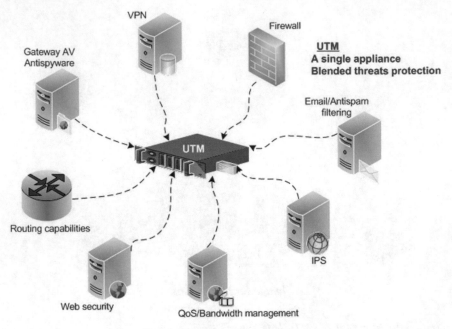

Image Src: dttstores.com

Web application firewalls:

A Web Application Firewall (WAF) protects the web apps by tracking the HTTP traffic and filtering it, which flows in between a web application and internet. WAP handles the threats such as SQL injection, cross-site-scripting (XSS), cross-site forgery and file inclusion. A web application firewall cannot safeguard the system from all sorts of attacks though it is situated in the protocol layer 7. Usually, this security advancement is basically a set of approaches that collectively establish a comprehensive security, countering a varied number of network attacks in [14]. A firewall between the internet and the web application is established when a WAF is installed at the front of the web application. Generally, a client's computer privacy is protected by a proxy server by using a proxy, whereas WAF protects the server by using a reverse proxy from disclosure of data. A WAF functions by a series of guidelines that are sometimes referred to as policies. Via filtering out harmful traffic, these policies help to defend against vulnerabilities in the framework. The benefit of a WAF is partially due to the ease and speed at which policy adjustment can be applied, providing a quicker response to different attack vectors; rate restriction can be easily implemented by changing WAF policies during a DDoS attack [14].

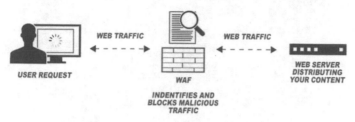

Image Src: quoracdn.net

Next-Generation Firewall (NGFW):

This is more like an advertising word that became famous among firewall manufacturers lately, says *Harris Andrea, 2020* in [15]. Basically, an NGFW incorporates nearly all the forms which we have mentioned above into one package. It is a Stateful firewall of hardware that also offers security and analysis at the application level. This form provides deep-packet filtering and is good at detecting malicious traffic in all OSI model layers. A NGFW generally offers advanced detection/prevention of intrusion, functionality, antivirus, monitoring of the program, etc. These are often individually licensed and additional payments have to be charged by the customer to allow some/all of the security. To collect threat-intelligence information from the cloud, some NGFWs connect with a manufacturer's cloud protection service (e.g. Fortinet FortiGuard, Cisco Talos, etc.).

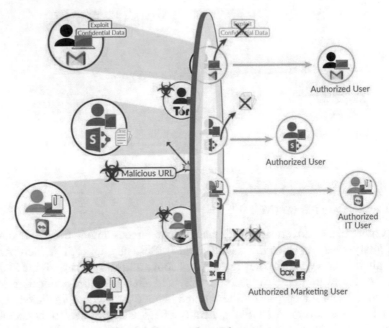

Image Src: technotification.com

5 Next Generation Firewall NGFW

In [16] *Alfonso Barreiro, 2012* says, the word "Next Generation Firewall" (NGFW) is used to define technologies that, by incorporating security capabilities such as intrusion prevention, go beyond standard firewall functions. The idea and market segment is relatively recent, as Palo Alto Networks was widely credited with producing the first of these sorts of products. Gartner has mostly advocated the concept outside of security vendors, using it to describe products with the following capabilities:

- 'Integrated' network intrusion prevention.
- Bridged and routed modes.
- The ability to use sources of external intelligence.
- Identity awareness in both group and user control.
- Features of standard firewalls such as network address translation, packet filtering and VPN capabilities.
- The developing ability and using ability of 'extra firewall' information to strengthen the decisions of blocking, such as using credibility services or identification services including the Active Directory.
- An "application awareness" that can classify applications and implement controls at the application layer (like allowing calls on Skype but restricts file transfers).

Firewalls also act like a border between the Internet and the private network of an organization. Most of the standard firewalls provide the safety focused at monitoring particular ports, their protocols and limiting the inward & outward flow of traffic with respect to individual IP addresses. Nonetheless, we can find mostly web-based attacks these days, which passes via https (port 443) & http (port 80), since most firewalls do not detect malicious software or traffic passing through these ports. To effectively protect against such attacks, the trusted firewall must evolve says *Alfonso Barreiro, 2012* in [16].

According to Eric Geier, 2011 in [17], in the management aspects of security as well as bandwidth, the fine-tuning of traffic generated by NGFWs will support. They have the ability to detect more malicious activity, because they are smarter and have deeper inspection. They also provide Quality of Service (QoS) functions and serve as content filters, therefore, the priority of higher bandwidth depends goes to the applications which has higher priority. The growth of cloud computing and outsourced software as a service (SaaS) suppliers resulted in high demand for NGFWs, likewise it is needed for overall security in general.

Through supplying it with the capacity to comprehend the specifics of the web application traffic going through it and take steps to block traffic that may compromise vulnerabilities, NGFWs provide added meaning to the decision-making phase of the firewall. As they can be combined with security intelligence systems, NGFWs are also well suited to counter emerging persistent threats (APTs). NGFWs can be a low-cost alternative for the businesses which pursues for enhancing the simple device supply, by using the inspection facilities, application awareness, safety mechanisms and awareness software. says *Margaret Rouse, 2018* in [18].

A next-generation firewall (NGFW) delivers features above a state-of-the-art network firewall, initially developed by Check Point Software Technologies in 1994 in [19]. A Stateful firewall is a network security system which filters the network communication such as inward and outward of network traffic, based on the IP addresses and Internet Protocol (IP) port. The payload of certain packets is intelligently analyzed by the NGFW in order to connect the new link requests with current links that are valid. Additional functionalities like device monitoring, automated intrusion prevention (IPS), and even more sophisticated vulnerability prevention technologies like sandboxing are applied to a next-generation firewall in [19].

Extending the functionalities a network firewall usually has, the above NGFW features include high-availability capabilities, dynamic support for routing protocols and network address translation (NAT). *Alfonso Barreiro, 2012* says in [16] the term UTM (Unified Threat Management, invented by IDC) is often somewhat ambiguous and represents a multipurpose monitoring system outside the typical firewall. A supplier could provide devices in various segments using either a description or categorizing them. UTM is the most common segmentation technique to refer to devices targeted at small to medium-sized enterprises, and the term NGFW is specifically used for devices designed for larger businesses. There is a rather vague difference when comparing a Unified Threat Management (UTM) system with a next-generation network firewall. UTM devices are specifically developed considering the consumer segment of businesses such as small to medium, where the security solution is robust. Over standard firewalls, such as antivirus, anti-spam, or also intrusion detection schemes, UTM products typically provide additional functions (IPS) [19].

Steven Rainess, 2016 explains in [20] that in critical areas of the network, additional functionalities such as malware protection and deep-packet inspection are commonly deployed. In their catalog, businesses such as Palo Alto, Forcepoint, Fortinet, and Cisco all have Next Gen Firewalls. However, they don't always provide the same functionality. Be mindful that only because a vendor uses the phrase, it does not always mean that this specific set of functions would be given by their product. Vendors continue to improve their products, play to their strong points or incorporate talents to separate themselves from the competitors. Maintenance of these firewalls can sometimes be a massive undertaking; therefore you want to confirm that the interface and the process of modifying policies and activities are convenient with you. You may not even have to manually upgrade each firewall if you have several firewalls in place when a change has to be made. If possible, you need to be able to centrally monitor it and be able to quickly access and customize it. Any products, which may or may not be a concern, may be handled from the cloud.

Characteristics of a Next Generation Firewall:

According to Eric Geier, 2011 in [17] the following are some features of most of the Next Generation Firewalls:

Standard firewall features: They provide standard (first-generation) firewall technologies such as VPN, network address translation (NAT) and stateful port/protocol inspection.

Intrusion prevention: They will also be able to conduct intrusion detection and avoidance, becoming more insightful also with tighter traffic inspection. Any next-gen firewalls may have ample IPS functionality that would not require a stand-alone IPS.

Directory integration: Most NGFWs provide support for directories (i.e., Active Directory). For e.g., handling of approved applications which are dependent on users and classes of users.

Malware filtering: To ban applications that have a poor reputation, NGFWs may also have reputation-based filtering. This will probably search virus, phishing, and other malware pages and apps.

SSL and SSH inspection: SSL and SSH secured traffic can also be reviewed by NGFWs. They will decrypt the traffic, ensure that it is an approved program and review other rules, and then re-encrypt it. This offers extra protection against malicious programs and operations that aim to mask encryption and keep the firewall from being used.

Application identification and filtering: This is the principal quality of NGFWs. Instead of only opening ports for any and all traffic, they will classify and process traffic depending on the individual software. This prohibits non-standard ports from using malicious software and actions to bypass the firewall.

Defining the Need of Next Generation Firewall:

Steven Rainess, 2016 suggests in [20] to make sure you select one that matches your needs when looking for a next-generation firewall. Some solutions, for example, may have CASB security, while others may only be able to exploit SD-Wan technologies. However, they all should be able to include IPS (Intrusion Protection System) and perhaps some Anti-Malware functionality. Any of the NFFW offerings can be subscription-based, such as premises Antivirus products, thus this must be taken into consideration when budgeting. In order to provide up-to-date identification and prevention along with remediation, others can actually engage with your onsite AV product. Conceivably, this could work in two separate directions. For example, if the NGFW knows that a threat has been detected and blocked by other NGFWs, it will alert all endpoints on the network to automatically quarantine this threat. This could also occur in some other case, but if an infected file is picked up by the endpoints, it might warn the firewall(s) to not let this threat cross the network. To have a well-rounded approach, many vendors actually work with other vendors, hence this must be properly considered. In their portfolio, vendors would still have other items that may be of importance and operate specifically with their NGFW deals as well. Any of the other items most likely to be available are DDOS, switching, DLP and email/web defense.

Benefits of Using A Next Generation Firewall:

The principal benefit of an NGFW is the potential to allow Internet applications to be used safely, allowing increased efficiency to users when restricting non-suitable applications. Next generation firewalls (NGFW) accomplish it through deep packet filtering in order to recognize, monitor programs regardless of which IP port is used by the device. A network firewall which is deployed at the organization's perimeter has a standard security policy to block connections those are inbound but enable connections which are outbound. It is possible to apply some limits, although outbound Web traffic is permitted in general. Customers get a seamless user experience provided by the applications using Web port 80, which is an accessible open port to the internet. This applies to the applications that are accurate in allowing the team to work more effectively and the least preferred applications according to company's interests. The next generation firewalls give businesses so much exposure about what applications their executives use and regulate about the use of their applications in [19].

According to *Chris Brook, 2020* in [21] the distinguishing characteristics of the Next generation firewalls create some unique advantages for the organizations those uses them. NGFWs can block the entry of malware into a network, something which would not be accomplished by the traditional firewalls. Advanced Persistent Attacks (APTs) are well suited to handle them. For businesses looking to improve their basic security, NGFWs can be a low-cost choice since they can combine the work of firewalls, anti-viruses and other security applications into one approach. Inspection services, application recognition and also a safety mechanism and awareness tool are the features of this that supports the offering at all odds.

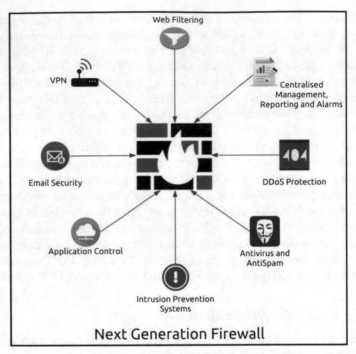

Image Src: firewall.firm

How do Next Generation Firewalls implement User Control?

At the very least, a network firewall security policy rule states that a connection is either permitted or rejected which is initiated from a specific source to a specific destination. Traditionally, in [19] the source point and the destination end are identified as an IP address allocated to a system such as computer or a laptop or a wider network address like an organization that contain several clients and servers. This concept of static address policy is hard to understand, but it also does not work well for people who have multiple IP addresses to set security policy when they roam around the organization and while operating off-site. Through interacting with third-party's client repositories such as Microsoft Active Directory, next-generation network firewall vendors overcome this. Dynamic, identity-based policies provide people, groups and devices with granular access and power and are easier to handle the policies which are IP-based and static. Unified console managers identify artifacts only once in a single console. The IP is assigned to an individual and community through verifying a user directory (third party) when the network firewalls finds an initial communication. This complex IP mapping consumer frees administrators from modifying the security protocol continuously [19].

How do Next Generation Firewalls enforce Threat Prevention?

Attack protection features can be considered as a logical extension of the capabilities of deep-packet scrutiny for next-gen firewalls. They often check the traffic for suspected exploits to current vulnerabilities when the traffic moves via the network firewall unit (IPS). To detect malicious activity, files may be transferred from the computer to be emulated in a simulated sandbox (sandbox security) [19].

Importance Of Next Generation Firewalls:

According to Chris Brook, 2020 in [21] for any organization, installing a firewall is a necessity. Getting a next-generation firewall is just as critical in today's environment. Every day, risks to personal computers and bigger networks are evolving. With the versatility of an NGFW, it defends computers and businesses from a much wider range of intrusions. While these firewalls are not the best option for any organization, the advantages that NGFWs can offer must be carefully weighed by security experts, since they have a very significant upside.

Things to consider when evaluating Next Generation Firewalls:

Alfonso Barreiro, 2012 in [16] says Next Generation Firewalls are dynamic machines, regardless of the terms, and the absence of a specification will make it very difficult to compare apple-to-apple measurements of different items. You must have a detailed view of the needs of the company and do rigorous research to assess if a system with a certain collection of capabilities will assist you.

Architecture: On a single inspection, next-generation systems can implement all their security capability, showing true convergence of all their components instead of merely bundling various product engines into a single package. A lack of convergence may also suggest that there could be trade-offs in security capability to compensate for inefficiencies or retain an acceptable level of efficiency (for example, it may have decreased IPS detection capabilities).

Ease of use: The prospect of reducing the difficulty of handling diverse security items is a significant factor for the acceptance of these systems. This can be expressed by the management framework, ideally being convenient to use and having the ability to effectively identify rules or regulations that are as granular or complicated as desired. When configuring various functionalities on a computer, particularly if their integration is missing, it is not uncommon to notice unexpected variations in the interface. These modern forms of firewalls, as with most protection devices, are not perfect solutions. In both the original setup and its continuing maintenance, implementing these takes a lot of effort. However, a good implementation could potentially help boost your odds against the current wave of network threats [16].

Throughput performance: All of the additional features, tests and inspections done by these machines in [16] will undoubtedly serve as a speed hump to the traffic flow. Try to ensure that the throughput follows the requirements for your production environment after all the safety features have been allowed. It is also taken into account in the assessments that the amount and sophistication of the device's policies or regulations will also be a consideration that can influence its overall performance.

Comparison between Traditional Firewall and Next-Generation Firewall:

In [22] *Alex Lyu,* defines that traditional and next-generation are the two most basic definitions for enterprise-level firewalls. The more sophisticated of the two forms is NGFW, the next-generation firewall. For your company, it would provide the most robust security. But what are the differences between the traditional firewall and the Next-generation firewall? How can your organization benefit from the differences? All of them have the same basic purpose, but unique systems, specifications, capacities and difficulty levels can differ enormously.

Traditional Firewall:

Standard/traditional firewalls are particularly designed for traffic monitoring. The inward and outward of the traffic of a network is managed basing considering the source IP address, protocol, port and destination IP address. If we speak regarding traditional firewalls, we talk about traditional state-of-the-art firewalls. The characteristics of the standard firewall in [22] are confined as follows:

- When users or devices enter a shared or unreliable network, the Virtual Private Network offers a safe access to the network.
- Stateful inspection, Instead of isolated packets, traffic is characterized by flow. Policies should be adjusted to the flow of traffic, and decisions about actions can be made.
- The network administrator is assisted by packet filtering to ensure that all the entry and exit of traffic is under inspection. Firewalls may terminate the communications those are initialized on behalf of users by a suspect source.

Stateless or Stateful?

Stateless inspection in [23] means that only each packet can be independently reviewed by the firewall, and it is unable to distinguish the larger meaning. Most of the traditional firewalls work only at the level of stateless (or 'state-unaware').

Stateful monitoring features are available for the more advanced conventional firewalls, which ensure that they can detect the operating state of packets attempting to reach the network. They are 'state-aware' in other words, and can differentiate between the secure, the potentially harmful and the outright malicious. NGFWs go a step further than regular Stateful inspection which we'll discuss further down.

Next-Generation Firewall:

The above mentioned and so much else is achieved through a next-generation firewall says Alex Lyu, 2020 in [22]. In conjunction to access control, guidelines to obstruct modern threats are given in a more granular manner. A next generation firewall must have the following according to Gartner's description:

- Standard capabilities of firewalls such as Stateful inspection
- Techniques for resolving emerging threats to defense
- Upgrade routes to provide potential feeds for details
- Integrated intrusion prevention

- Threat intelligence sources
- Application awareness and control in order to detect and restrict/block any risky apps.

There are 4 main benefits to be delivered by the next-generation firewalls to the organization:

Application awareness and control:

According to *Mechsystems, 2020* in [24] 'Application awareness/recognition' is the largest difference between a Next-generation firewall and a traditional firewall. Traditionally, firewalls have relied on traditional ports for attack tracking. NGFWs track the traffic from layers 2–7, and then decide whether the transmitted traffic is malicious or anything else. The Next-generation firewall can allow for application control since it is being application aware. The end user identity will be tracked by NGFW usually by means of the Active Directory os something similar. The IT team will put usage controls in place, based on both the program and the user, to regulate the traffic arriving and leaving the network and how/what the individual can submit and receive. Facebook may be permitted, for instance, but videos are prohibited from being played. This method of system could also decrease the requirements of the bandwidth, and save many vital resources to be used somewhere else [24].

Intrusion prevention system:

Intrusion prevention framework contains the potential to effectively track and prevent intrusions. The monitoring will apply to a dynamically modified cloud database against even a zero-day attack says *Alex Lyu, 2020* in [22]. A NGFW can restrict traffic only to approved applications, alleviating the hazards of obscure applications. A great deal of the usage of the bandwidth can be cut down by this. Threat Analysis provides the latest intelligence for the firewall and other monitoring equipment to identify and deter new threats.

Simplified Infrastructure:

According to *Mechsystems, 2020* in [24] a next-generation firewall is like an all-in-one kind of solution. Among other security, the best offerings come loaded with anti-virus, spam filtering and deep packet inspection. Instead of needing to keep jumping on and upgrading new devices, IT will handle one device.

Guaranteed Speed:

The speed flowing through the network can be throttled by conventional firewalls. The more security services are introduced, the more pace sinks into the firewall as it heads out to the end users. No matter how much security is allowed, the throughput on an NGFW does not improve. The pace that comes in goes immediately straight out explains *Mechsystems, 2020* in [24].

Deep packet inspection (DPI) guarantees that the contents of the packet are carefully reviewed, while regular packet inspection examines only the header. NGFW will take care of a single packet's complete context says *Alex Lyu in* [22]. Whereas, in

[24] *Mechsystems* suggests, a Next-generation firewall facilitates scheduled person-alized user activity reports and the ability to export them to a particular format (e.g. HTML, PDF, Excel). IT will access a live view of the web operation of users and allow immediate alerts for violations of policies and network outages. This is crucial for ensuring where urgent danger notifications will help defend consumers from damage.

Similarities between the two:

According to *Omninet, 2015* in [25] the basic purpose of both standard firewalls and NGFWs is clearly the same: to secure the network and data properties of an enterprise. Both of them have a combination of the following in terms of the device components bundled by the two:

- Stateful inspection or dynamic packet filtering, which tests the legitimacy of a link on every firewall interface.
- Translation of port address, which makes it possible to map several devices on a LAN to a single IP address.
- Support for a virtual private network (VPN) that retains the same access and protection characteristics as a secure network across a section of the link that flows across the internet or other similar system.
- Translation of the network address for re-mapping the IP addresses used in packet headers.
- Static packet filtering that filters packets based on protocols, ports or addresses at the interface point of a network [25].

Differences between the two:

The following are the extra security features of Next-generation firewall stated *by Omninet, 2015* in [25] which are usually not found in the traditional firewalls:

Non-disruptive, in-line configuration of bump-in-the-wire (BITW), in which 'stealth' firewall exists within the subnet so that it would filter the traffic between hosts.

- Integrated signature-based prevention scheme for intrusion (IPS) that defines the types of attacks to be screened and recorded.
- Identification of packages/applications with the help of payload analysis, header inspection and pre-described software signatures, plus enforcement of network protection coverage on the software level, due to the fact that packages/applications (in preference to networking services and components) have turn out to be the finest region of exploitation nowadays with the aid of using malware and similar different attacks.
- Ability to integrate outside firewall details, along with black lists, directory-based rules and white lists.
- Decryption of the Secure Socket Layer (SSL) to recognize unacceptable encrypted programs
- Upgrade route to provide possible security risks and data feeds.
- Granular management of applications, or highly thorough control of them.
- Absolute visibility of the stack, which directly correlates with the applications control [25].

In [26] the difference between the traditional/standard firewall and the next generation firewall given by *Rashmi Bhardwaj, 2020* is shared in the following table:

PARAMETER	TRADITIONAL FIREWALL	NEXT GEN FIREWALL (NGFW)
Application Visibility and Application Control	Partial	Detailed
CAPEX and OPEX (considering all feature requirement)	Higher since separately need to buy and maintain	Considerable reduction since all services will be bundled into single box
IPS (Intrusion Prevention System)	Not Supported	Supported
NAT	Supported	Supported
Reputation and identity services	Not Supported	Supported
Traffic filtering (Port, IP Address and protocol based)	Supported	Supported
VPN	Supported	Supported
Application level awareness	Not Supported	Supported
Working Layer	Layer 2 to Layer 4	Layer 2 upto Layer 7
Throughput and performance	Lower than NGFW and drastically reduces when additional services introduced.	Much higher than traditional Firewall and doesn't change much on introduction of additional services.
Reporting	Standard reports	Customized reporting upto user level giving near real time detail with plenty of additional reporting options like download format etc.

Img Src: ipwithease.com

Benefits of a Next-Generation Firewall over Traditional Firewall:

According to *Deren Chen, 2019* in [27] the Benefits of the Next-Generation Firewalls over the traditional Firewalls are as follows:

Next-Generation Firewalls Provide Resource Efficiencies:

A next-generation firewall's primary benefit is the sophisticated detection technology that these strategies add to the table. The threat environment is continuously evolving, and to detect and deter unknown malicious malware from slipping into a network, an NGFW should exploit threat intelligence information. In particular, various security solutions are integrated on a single architecture by NGFWs, including web security capabilities, application visibility and intrusion prevention [27].

Next-Generation Firewalls Are Often Cost-Efficient in the Long Run:

The department actually reduced its expenses from a financial point of view. While switching to the NGFW was more costly than merely removing the legacy firewall technologies, having an NGFW permitted the agency to substitute existing protection devices with a single framework. The Department substituted a single product capable of satisfying all three requirements for its firewall, intrusion protection system and network protection solution. On all three options, the NGFW enhancement expense was much smaller than the agency's combined expected replacement costs [27].

Next-Generation Firewalls Provide Resource Efficiencies:

However, the benefits didn't stop with such direct expenditures. The technology unit at this agency, like many government departments, was severely understaffed and lacking the specialists required for managing and tracking current security technology sufficiently. Centralizing security technologies allowed the department to simplify management roles and significantly enhance the productivity of employees. It also provided the IT specialists of the organization more insight about how their organization uses bandwidth in [27].

What's next for Next Generation Firewalls?

As security concerns continue to evolve, enterprises are turning away from the Next Generation Firewalls and forwarding towards a new "Network Firewall" technology referred by Gartner. Network Firewalls, along with external security functions all over the endpoint, web, server, data center, and IoT, offer real-time threat intelligence. A firewall is an integral aspect of the security infrastructure of any enterprise and can help secure sensitive data, comply with regulatory standards, and direct the companies to achieve digital transformation [19].

6   Adaptive Firewall with the Help of AI

According to *Collin Rauen (2019)* in [28] a traditional firewall focuses only on the prevention of a threat, when an attack on their defenses is detected. Adaptive firewall does not have this problem, since it uses a threat-basis approach with an in-depth protection and intelligence throughout the attack phase, that gives the control over the complete attack sequence, i.e., before the attack, during and after the attack. Administrators can monitor on what has happened and can take enough measures to alleviate the problem. Furthermore, it can identify and protect against the unknown threats in [28] by incorporating few integrated layers of security like,

- Firewall routing and switching
- Identity-policy control and VPN
- Intrusion prevention
- URL filtering.

The above mentioned features work correspondingly to identify the emerging threats which are usually missed by the traditional firewall's security. In standard packet filtering, the traffic is limited between the network addresses of source and destination and allows only particular applications to travel across the network. Even though router has the static filtering, the rules of the filtering are framed in a specific order. These rules cannot adapt to the unforeseen circumstances and can leave serious vulnerabilities. This problem is solved by the adaptive firewall, because it adapts to particular traffic on the network and controls the connections by opening, locking and limiting them according to the demand between the recognized hosts for approved applications.

A proper configuration gives a solid control over the firewall regardless of the location, considering there is an access to the internet with a supported web browser such as, Internet explorer in case of Windows operating system and Netscape in case of Linux operating system. Along with tracking the access attempts, adaptive firewall also delivers the real-time status indication of the network traffic which is both outbound and inbound [28].

Functioning of AI embedded adaptive firewalls:

The adaptive firewall mostly works on the basis of collection of data and analyzing/evaluating it. This evaluation is done by the AI which is connected to the sandbox of the firewall, as well as connected locally in the organization and in the cloud. For this to be done, the firewall should be provided with information through a 'feeder' as well as lets it to collect information from the internet, which helps in studying the behavior of the programs and users by self learning. Using this shared/collected information, the AI can detect any anomalies that occur in the system, say it in personal computer or in an organization. However, the firewall has some static rules in order set the regulations according to the user, whether the user belongs to a particular company or he is using the firewall for home network. According to the requirement, the rules may vary, but the behavior is always same like, collecting the data, alerting the user when an anomaly occurs, acquiring the information from the feeder etc. Figure 1 gives the idea of how the information is provided or collected by the firewall.

From the above figure, we can see that the firewall is connected to various entities, where it gets the signature of different attacks or viruses and updated attack vectors through the vendor via feeder which is connected to the cloud, where as studies the behavioral pattern according to the end user's choices and actions. Also, parallely collecting information about the attacks those are happening around the world through 'World Wide Web'.

Functioning of Adaptive Firewall: Fig. 2 shows an example of how an adaptive firewall responds according to anomalies.

If we observe the above figure, we can see that the firewall receives data from various sources like, an e-mail from a website, the cache from the browsing data through the internet and even an unauthorized access from an attacker. While the internet data is filtered and allowed to reach the system of user/administrator, both the e-mail which has a malicious content as well as the unauthorized attempt was blocked. Whereas regarding the unauthorized access, the AI in the firewall sends an alert to the user/administrator as well to the cloud which consequently alerts the

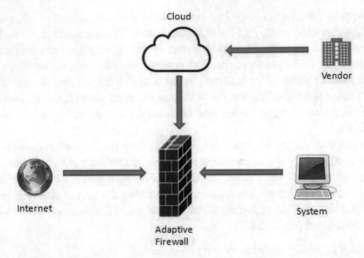

Fig. 1 This figure shows the design of how information is collected by the firewall

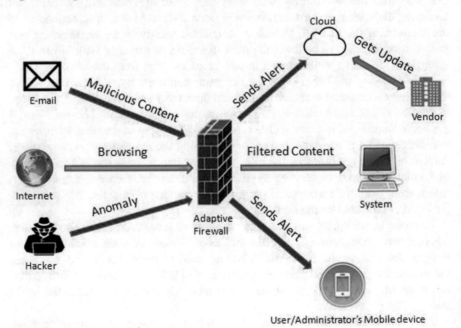

Fig. 2 This figure shows the adaptive firewall methodology

vendor stating about the unusual activity and asks for permission to proceed if it is done by the very user/administrator itself. Since, it is an attempt of an attacker (in our example), the user can simply deny it and the firewall restricts the access from that respective channel. After this has happened, this activity is registered in the cloud and is kept in the logs for future references.

In another example, when a user tries of open a website which has malicious content that can be harmful for the system, the firewall alerts the user about the risk and informs the consequence if the user still wants to proceed. If the user proceeds despite the warning, the AI in the cloud alerts this decision of the user to the vendor. However, since it is an individual decision of the user, the vendor does not change the rules to allow such content for others and the firewall gives the same warning to any other user who tries to access the same website.

Emulator Environment: The firewall once installed, provides an emulator, where a user can access the internet and other options via a firewall shell, where it is connected to the cloud and updated continuously by the vendor. Hence, whenever there is an anomaly, the AI gets alerted and takes appropriate action. The following Fig. 3 gives the idea of how the emulator sets up a system in the safe environment.

From the above figure, we can see that the system is connected to the emulator via browser and the mobile is connected by means of an app specifically designed for emulator. Through this, the user can access the internet or any other programs in the respective system. An internet connection is mandatory while connecting to the emulator as it is constantly linked with the cloud.

Objectives:

- Strengthen the restrictive control with the help of Artificial Intelligence (AI) to avoid cyber threats.
- Dedicated monitoring of the anomalies in the network.
- Restriction on Network services like Directory, File services and e-mail services etc. to only limited staff.
- Intense screening of the incoming and outgoing traffic in order to filter the vulnerable data.
- Advanced pre-defined set of network protocols for more effective inspection capability.
- Using 'Cloud-Login' as second factor to make the authentication stronger.

Requirements:

Though the firewall is adaptive in nature and has the pre-defined methodologies of previous attacks, it still needs additional information to be fed in order to learn and analyze the behavior of unusual activities, because, there are always new form of attacks and viruses crop up. For this information sharing to happen, a feeder needs to be connected to the firewall that supplies the information continuously, which makes the firewall aware of the new attacks and the techniques and stays updated. This feeder is situated locally at the vendor's end. Through this, the firewall learns continuously by means of a feeder as well as through self-learning. Internally, the AI gets the information from the feeder and externally, the AI takes the information from

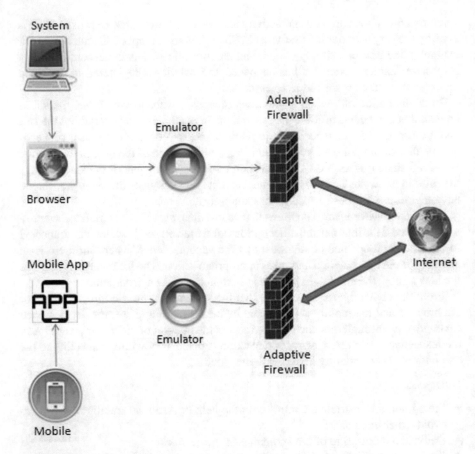

Fig. 3 This figure shows the adaptive firewall methodology

the internet, which means, the AI keeps itself updated according to the current trend that's followed world-wide, like the new attacks being recorded and the remedies for such attacks. After the 'feeding of information' is taken care, the vendor needs to work on the set up that performs the 'notification' part, where the AI notifies the vendor and the organization or an individual user by multi-factor authentication in order to proceed for the next step. For instance, if at all a DDoS attack is about to happen, then the AI detects the unusual load of requests and alerts the client to his email or mobile phone stating/alerting about the unexpected number of requests. And if the user/administrator permits the activity, then the AI allows the request but alerts the vendor about the permission given by the user/administrator for the anomaly. The simplified requirements for constructing the adaptive firewall are as follows:

- Information sharing
- Feeder
- Cloud authentication

- Multi-Factor Authentication (hybrid authentication)
- Pre-defined methodologies of previous attacks
- Continuous learning (Internal & External)
- Combination of Feeding and Self-learning.

Challenges in implementing the design:

Since this is a new design, there are high chances that we may face number of challenges. Hardware and software challenges are the most likely to rise up first. The vendor should explain the necessity of the budget the organization or an individual has to invest in this new model of firewall, because, the cost of making the design and maintenance varies compared to the traditional firewalls. If these are the challenges we may face after the software is ready to incorporate, there are other challenges that the vendor may face before the software is made, such as, the approval of the design, because, the involvement of the artificial intelligence makes the concept complicated. The behavior sometimes may go out of hands since it takes decisions itself while also following the given set of rules, which means, not all the decisions that the AI makes might be rational. And that is why, during the design of the AI itself, it has to coded in a way that it is restricted to some level of self decisions. And then comes the integration challenge, such as, connecting the firewall to the local servers through software and hardware respectively, synchronizing the firewall with the cloud and so on. And then comes the deployment stage where the 'Research and Development' department can find any faults and bugs in both the software and hardware environment. After the bugs and faults are take care of, the firewall is kept to test in an actual controlled environment. During this stage, it is essential that the software should not arise any bug, because if so, the whole process needs to be re-examined which might consumes enormous time, money and effort. The above discussed points can be streamlined as follows:

- Convincing the organization in terms of software
- Convincing the organization in terms of budget
- Design approval
- Integration challenge (software)
- Integration challenge (Hardware)
- Deployment challenge (hybrid/on premises)
- Deployment challenge (Cloud)
- Testing challenge.

Risks:

People's side: Resistance to buy this concept. This applies to both organization and an individual, because, the primary reason would be the cost which would be way higher when compared to the traditional firewalls. That's because of the design the firewall is made as well as for the maintenance to keep the firewall up-to-date and running efficiently. And then comes the trust part, where most of the people feel resistant to try new software because they are unaware of the bugs that might occur and many organizations cannot wait for the patch-up to release since the information related to the company is vital and they cannot take risk on it. Hence, the vendor needs to convince the organization as well as the individual user about the security of the personal/company's information and if possible the vendor should demonstrate the working of the firewall to build-up confidence on the software.

Process Side:

The only risk assumed to be existing in the process of adaptive firewall is, the 'offline' access. When the user/administrator works offline or if the internet connection is failed abruptly, then the emulator stops working as it is connected to the cloud via internet. The emulator is the medium that lets a user/administrator to work in a safer environment, which can be accessed through the respective device by means of an internet connection. The AI is active in this shell, and if the system looses the access to this shell, then the firewall is going to work in a traditional/standard way, however works efficiently with the updates installed until then. Therefore, internet connection is a mandatory for any system in which the adaptive firewall is incorporated.

Technology Side:

The presumed risk on the technology side would be the compatibility issues of the emulator on different devices. Though this emulator is designed to work in almost all the devices, there are chances that it might not work on all platforms. Some devices needs to be updated according to the latest operating system, other devices like mobile phones may have mid-end processor that could not handle the emulator efficiently. If this can be solved, the risk on the technology side is almost straightened out.

Distinguish adaptive nature:

The adaptive nature of the firewall can be discussed in detail in the following features:

Authentication: The authentication is a feature which the firewall alerts the user/administrator to allow the accessibility, authenticate to run a program. Here the firewall determines and understand the following.

a.   The required way of accessing.
b.   Analyze the pattern and way of access.
c.   Understands the login pattern, Login frequency, creating the required rules as per questioner, allow to run various program/service running, allow the ports to open, correspondingly analyze the outward and inward traffics on the network layer.
d.   Adaptive Intelligence (AI) includes two types of authentications in 'Adaptive firewall', namely, 'Two-Factor Authentication' and 'Cloud Authentication'.
e.   Based on the variation noted, agreed method of alert mode – detection, prevention modes will be applied respectively.

Two-Factor Authentication: When an anomaly is detected by the AI, the firewall sends a message to the email and mobile phone of the administrator/user to confirm if that change is done by the administrator/user itself.

Cloud Authentication: When an unusual behavior occurs in the system or someone tries to access the system in an unauthorized way, the firewall sends an alert to the cloud where the user/administrator as well as the vendor is alerted about the change.

Authorization: The authorization is the process of allowing the program or in simple words, overriding the firewall's advice to run a specific program/service by the user. The authorization can be given by the administrator or by the firewall itself. The administrator overrides/allows the new anomaly knowing the risk, where as the firewall authorizes the program to run, since the firewall learnt the program/anomaly was safe to run by learning through previous behavior. Authorization includes the accessible permissions, required and allowed ports, newly identified exceptions, over ride executions also.

Accountability: Adaptive firewall performs audits of the activities, i.e., it maintains the logs for evidences. This helps in tracking end-to-end activity in a system. For example, basing on these logs, the firewall decides which programs/services are to be allowed even though they are suspicious in nature because the user might have allowed it previously and the log is created that user have permitted. The firewall saves these actions for future references. Accountability, including activity logs of user access, traffic passed, noted exceptions, are also analyzed to provide health check kind of report, so the end user will be able to visualize and do the needful. This way the core technical data are analyzed and simplified for end user usability.

Prevention versus Detection: Usually traditional firewalls detect the malicious content in the traffic and tries to prevent it from entering into the system but sometimes that content is somehow bypassed by the hackers by exploiting the vulnerabilities. But in the Adaptive firewall, when the anomaly is detected, the firewall alerts the user and if the user takes an action on the anomaly, the firewall remembers the action and if the same activity happens for the next time, instead of alerting the user, the firewall forbids the action since it already knew that the user is going to reject it. Hence, the adaptive firewall is detective in nature and also protective.

AI-Rules built capacity:

The device will seek 'do it yourself' (DIY) based approach, allowing any and every user to interactively approach the basic necessity. Hence based on the end user response the initial set of rules are set up and also best practices are always recommended at each step. Thus, dynamic way of results can be achieved.

The AI, allows the end users to also share their Meta data that is used to analyze and appropriately suggest the required enhancement periodically. Further based on worldwide patterns the new suggestive ways will be shared for necessary considera-tion. The adaptive kind includes understanding the user requirements, suggest them with the newer requirements including new policy building, considering proactive steps like revisiting the required rule set, auto detecting the changes like new device addition, new user access, new services opened, new traffic analyzed etc.

AI-Emulator:

The following are expected as part of the solution:

- The suggested solution will also be accessible via browser, hence will be able to access safely from anywhere around the World Wide Web.
- Post login the emulator will have the necessary space with apps also, that allows various programs used will be usable for the end users.
- As its emulator, any new installation will also be provisioned respectively. The solution allows provisions for the organization, home users thus profile based solution and other users can also be added.

Compliance with ISO 27001 Standards:

The methodology of Adaptive firewall technology is in compliance with various ISO 27001 standards in [29] such as.

A.5: Information security policies: Information security policies are the part of the policy that will be defined based on the questionnaire.

A.6: Organizing of information security:

A.6.1.1 Information Security Roles & Responsibilities: All information security responsibilities need to be defined and allocated. Information security responsibilities can be general (e.g. protecting information) and/or specific (e.g. the responsibility for granting a particular permission). Consideration should be given to the ownership of information assets or groups of assets when identifying responsibilities.

A.6.1.2 Segregation of Duties:

Conflicting duties and areas of responsibility must be segregated in order to reduce the opportunities for unauthorized or unintentional modification or misuse of any of the organization's data/assets.

A.6.1.3 Contact with Authorities:

Appropriate contacts with relevant authorities must be maintained. Remember when adapting this control to think about the legal responsibilities for contacting authorities such as the Police, the Information Commissioner's Office or other regulatory bodies e.g. around GDPR. Consider how that contact is to be made, by whom, under what circumstances, and the nature of the information to be provided.

A.6.1.4 Contact with Special Interest Groups:

Appropriate contacts with special interest groups or other specialist security forums and professional associations must also be maintained. It is important to understand the nature of each of these groups and for what purpose they have been set up (e.g. is there a commercial purpose behind it).

A.9: Access Control:

The firewall has access control policies which describe the authorization of network services access to users who are specifically authorized to use. Managing privileged access rights by restricting the access to unauthorized user when he tries to gain access to the privileged accounts. Removal/adjustment of access rights are reviewed from time to time as well.

A.10: Cryptography:

Cryptographic controls are developed and maintained for the protection of information. This is specifically a part of the solution design.

A.12; A.13 are the security measures for Communications and Operations; Network security respectively. The adaptive firewall provides security for the entire connected network by checking network controls, segregating the network services, making sure to follow the policies and procedures of the information transfer, monitoring the electronic messaging etc.

A.16: Incident Management:

Incident management is considered as one of the eminent feature of adaptive firewall because whenever there is an anomaly that hinders the information security, the AI immediately alerts the user/administrator, informing the vendor as well regarding the incident. Not only reporting but also responds to the incidence according to the documented procedures. After the incidence is reported, the evidence is collected and stored in order to prevent any future occurrence, hence self-learning.

A. 17: Information Security aspects of Business Continuity Management:

The planning, implementing, verifying, reviewing and evaluating of the information security measures helps in business continuity of the organization. This is achieved by continuous monitoring on the anomalies and reporting or alerting by the adaptive firewall [29].

7 Conclusion

To conclude, according to the current day scenario, one cannot predict how and when an attack takes place. No matter how many precautions are taken, the updated protection is being subjugated by a new attacking vector. Therefore, the importance of firewall is uttermost important to any system, either a home computer or an organization which has several computers connected to a network. The traditional firewall monitors the outflow and inflow of the traffic from the internet to the computer and avoids any unauthorized access and malicious content by means of a set of rules. However, these standard firewalls are upgraded time to time according to the latest trend of attack vectors yet the attacks are being taken place and the systems are being compromised, since these traditional firewalls can respond on a rule-based approach. Although Next-Generation Firewall (NGFW) has emerged as a cutting-edge firewall somewhere in 2010 by Palo Alto for the first time, according to *McCormack* in [30] the Next-Generation Firewall have some limitations like poor performance, complex set-up and ineffective deployment, despite being so effective in deep-packet filtering, intrusion prevention system (IPS), malware filtering, application identification and filtering. The proposed adaptive firewall has the features of Next generation firewall such as intrusion detection system, intrusion prevention system; deep-packet filtering and so on and additionally, the firewall can analyze the behavior of the traffic with the help of artificial intelligence (AI) that is connected to the sandbox. This adaptive firewall has self-learning capability as well as is fed by the vendor with the information about new attack vectors happening around the world, which helps the AI to analyze effectively when an anomaly occurs in the traffic/network. This can be customized according to the user requirement, such as, an individual user in home environment needs a different set up compared to an organization in which a network of computers has connected together. This is called 'do it yourself' (DIY) approach where the rules are set up according to end-user and also the firewall makes recommendations at each step. The user or administrator uses an emulator to access the apps or services, in which the environment is analyzed by the AI and sends the Meta data to the cloud. Whenever an anomaly occurs, the AI alerts the user or administrator be means of 'Two-factor Authentication' and the vendor by 'Cloud-Authentication', stating the unusual activity and asking permission to proceed further, and if the user or administrator still proceeds, then this choice is informed to the vendor. In this way, the AI in the firewall, not just obeys according to the given set of rules but also studies the behavior of the new requests comparing with the pre-defined methodologies and protects the system. No matter how much protective the system is, it always can be overridden by some other strong procedure, but the AI helps the user/administrator aware of the situation of anomaly before itself so that they can take action and avoid it at the starting stage itself. In this way, the system becomes 99% secured, which is hardly achieved by any other type of firewalls. Some works such as 'Designing an Adaptive firewall for enterprise cloud' in [31] by *Santosh Kumar Majhi, 2014 & Padmalochan Bera, 2014*; 'Development and investigation of adaptive firewall algorithm to protect the software-defined infrastructure of multi-cloud platforms' in [32] by *Irina Bolodurina, 2020, Denis Parfenov, 2020 & Vadim Torchin, 2020*; Policy And Implementation Of An Adaptive Firewall in [33] by *Theuns Venvoerd, 2020 & Ray Hunt, 2020* are few examples of Adaptive technology in firewall. However, all the above mentioned works focuses on handling a specific aspect, but the AI mentioned in 'Adaptive firewall with help of AI' works on a holistic approach that targets on detecting the errors or anomalies in an individual system or an organization, alerting the user/administrator and the vendor as well, preventing the known attacks and safeguarding the system.

References

1. Uky.edu (2021). http://www.uky.edu/~dsianita/390/firewall1.pdf. Accessed 19 January 2021
2. Ddine K (2021) Overview of firewalls: types and policies managing windows embedded firewall programmatically. https://www.researchgate.net/; https://www.researchgate.net/publication/315614367_Overview_Of_Firewalls_Types_And_Policies_Managing_Windows_Embedded_Firewall_Programmatically. Accessed 19 January 2021.

3. Byjus.com (2021). https://byjus.com/govt-exams/firewall-computer-network/. Accessed 19 January 2021.

4. eSecurityPlanet (2021) What Is a firewall? | Different types of firewalls. https://www.esecurity planet.com/networks/types-of-firewalls/. Accessed 19 January 2021.

5. Calyptix Security (2021) Types of firewalls: the good, the minimal, and the effective. https://www.calyptix.com/top-threats/types-of-firewalls-the-good-the-minimal-and-the-effective/. Accessed 19 January 2021.

6. Jain SR (2021) Firewalls and types of firewalls | information Security Blog. https://sanketrjain.com/types-of-firewall/. Accessed 19 January 2021

7. www.black-box.de (2021) 3507—types of firewall. Black Box. https://www.black-box.de/en-de/page/28180/Resources/Technical-Resources/Black-Box-Explains/security/types-of-fir ewall. Accessed 19 January 2021

8. Securitywing.com (2021) 7 different types of firewalls | securitywing. https://securitywing.com/types-of-firewall/. Accessed 19 January 2021

9. Dosal E (2021) What is a firewall? The different firewall types & architectures. Compuquip.com. https://www.compuquip.com/blog/the-different-types-of-firewall-architect ures. Accessed 19 January 2021

10. ComputerNetworkingNotes (2021) Types of firewall explained with functions and features. https://www.computernetworkingnotes.com/ccna-study-guide/types-of-firewall-exp lained-with-functions-and-features.html. Accessed 19 January 2021

11. Solid State Systems LLC (2021) Choosing a firewall: the most important features and types you need to know | solid state systems LLC. http://solidsystemsllc.com/choosing-a-firewall/. Accessed 19 January 2021

12. Services P (2021) What is a firewall? Cisco. https://www.cisco.com/c/en/us/products/security/firewalls/what-is-a-firewall.html#~types-of-firewalls. Accessed 19 January 2021

13. Firewalls T (2021) Types of firewall | 5 awseome types of firewall to know. EDUCBA. https://www.educba.com/types-of-firewall/. Accessed 19 January 2021

14. cloudflare (2021) What is a WAF? | web application firewall explained. https://www.cloudf lare.com/learning/ddos/glossary/web-application-firewall-waf/. Accessed 19 January 2021

15. Andrea H (2021) 7 types of firewalls. In: I.T and computer networks explained. Networkstraining. https://www.networkstraining.com/different-types-of-firewalls/. Accessed 19 January 2021

16. TechRepublic (2021) An overview of next generation firewalls. https://www.techrepublic.com/blog/it-security/an-overview-of-next-generation-firewalls/. Accessed 19 January 2021

17. Geier E (2021) Intro to next generation firewalls | esecurity planet. eSecurityPlanet. https://www.esecurityplanet.com/products/intro-to-next-generation-firewalls/. Accessed 19 January 2021

18. Rouse M (2021) What Is Next-Generation Firewall (NGFW)?—definition from whatis.com. SearchSecurity. https://searchsecurity.techtarget.com/definition/next-generation-firewall-NGFW. Accessed 19 January 2021

19. Check Point Software (2021) What Is Next Generation Firewall (NGFW) | check point software. https://www.checkpoint.com/cyber-hub/network-security/what-is-next-generation-firewall-ngfw/. Accessed 19 January 2021.

20. Rainess S (2021) Overview of next gen firewalls—need, sizing, and mangement. CCSI. https://www.ccsinet.com/blog/next-gen-firewalls/. Accessed 19 January 2021

21. Brook C (2021) What is a next generation firewall? learn about the differences between NGFW and traditional firewalls. Digit Guardian. https://digitalguardian.com/blog/what-next-genera tion-firewall-learn-about-differences-between-ngfw-and-traditional-firewalls#:~:text=A%20n ext%20generation%20firewall%20(NGFW,intelligence%20from%20outside%20the%20fire wall.%E2%80%9D. Accessed 19 January 2021

22. Lyu A (2021) Firewall: traditional vs next generation—sprint networks. Sprint Netw. https://www.sprintnetworks.com.au/blog/traditional-versus-next-generation-firewall/. Accessed 19 January 2021

23. Kedrosky E, Kedrosky E, Marketing S (2021) Next-Generation Firewall (NGFW) vs Traditional Firewall—security boulevard. Secur Blvd. https://securityboulevard.com/2019/06/next-genera tion-firewall-ngfw-vs-traditional-firewall/. Accessed 19 January 2021
24. M-Tech Systems (2021) Traditional vs Next-Generation Firewall (NGFW). https://www.mtechsystems.co.uk/latest-technology/traditional-vs-next-generation-firewall/. Accessed 19 January 2021
25. OmniNet (2021) Regular firewall vs next generation firewall | my digital shield. https://omn inet.io/traditional-firewalls-vs-next-generation-firewalls/. Accessed 19 January 2021
26. Bhardwaj R (2021) TRADITIONAL FIREWALL vs NEXT GEN FIREWALL (NGFW): detailed comparison—IP with ease. IP with ease. https://ipwithease.com/traditional-firewall-vs-next-generation-firewall/. Accessed 19 January 2021
27. Chen D (2021) Next-geneartion firewall vs traditional firewall: what's the difference? | CDW Solutions Blog. CDW Solutions Blog. https://blog.cdw.com/security/why-choose-a-next-gen eration-firewall. Accessed 19 January 2021
28. Rauen C (2021) Give your customers an adaptive security advantage | ingram micro imagine next. Ingram Micro Imag Next. https://imaginenext.ingrammicro.com/security/give-your-cus tomers-an-adaptive-security-advantage. Accessed 19 January 2021.
29. Controls I (2021) ISO 27001 Annex A controls—overview. ISMS.online. https://www.isms. online/iso-27001/annex-a-controls/. Accessed 4 February 2021
30. McCormack C (2021) The problem with next-gen firewall protection. Sophos News. https://news.sophos.com/en-us/2018/01/01/the-problem-with-next-gen-firewall-protection/. Accessed 19 January 2021
31. Majhi SK, Bera P (2014) Designing an adaptive firewall for enterprise cloud. In: 2014 international conference on parallel, distributed and grid computing, Solan, pp 202–208. https://doi.org/10.1109/PDGC.2014.7030742.
32. Bolodurina I, Parfenov D, Torchin V, Legashev L (2018) Development and investigation of multi-cloud platform network security algorithms based on the technology of virtualization network functions. The research work was funded by RFBR, according to the research projects No. 16–37-60086 mol_a_dk, 16–07-01004, 18–07-01446, 18–47-560016 and the president of the Russian Federation within the grant for state support of young Russian scientists (MK-1624.2017.9). In: International scientific and technical conference modern computer network technologies (MoNeTeC). Moscow, pp 1–7. https://doi.org/10.1109/MoNeTeC.2018.8572059
33. Verwoerd T, Hunt R (2002) Policy and implementation of an adaptive firewall. In: Proceedings 10th IEEE international conference on networks (ICON 2002). Towards Network Superiority (Cat. No.02EX588), Singapore, pp 434–439. https://doi.org/10.1109/ICON.2002.1033350.

Printed in the United States
by Baker & Taylor Publisher Services